Springer Theses

Recognizing Outstanding Ph.D. Research

For further volumes:
http://www.springer.com/series/8790

Aims and Scope

The series "Springer Theses" brings together a selection of the very best Ph.D. theses from around the world and across the physical sciences. Nominated and endorsed by two recognized specialists, each published volume has been selected for its scientific excellence and the high impact of its contents for the pertinent field of research. For greater accessibility to non-specialists, the published versions include an extended introduction, as well as a foreword by the student's supervisor explaining the special relevance of the work for the field. As a whole, the series will provide a valuable resource both for newcomers to the research fields described, and for other scientists seeking detailed background information on special questions. Finally, it provides an accredited documentation of the valuable contributions made by today's younger generation of scientists.

Theses are accepted into the series by invited nomination only and must fulfill all of the following criteria

- They must be written in good English.
- The topic should fall within the confines of Chemistry, Physics, Earth Sciences, Engineering and related interdisciplinary fields such as Materials, Nanoscience, Chemical Engineering, Complex Systems and Biophysics.
- The work reported in the thesis must represent a significant scientific advance.
- If the thesis includes previously published material, permission to reproduce this must be gained from the respective copyright holder.
- They must have been examined and passed during the 12 months prior to nomination.
- Each thesis should include a foreword by the supervisor outlining the significance of its content.
- The theses should have a clearly defined structure including an introduction accessible to scientists not expert in that particular field.

Wanai Li

Efficient Implementation of High-Order Accurate Numerical Methods on Unstructured Grids

Doctoral Thesis accepted by
Tsinghua University, Beijing, China

Springer

Author
Dr. Wanai Li
Tsinghua University
Beijing
China

Supervisor
Prof. Yu-Xin Ren
Tsinghua University
Beijing
China

ISSN 2190-5053 ISSN 2190-5061 (electronic)
ISBN 978-3-662-43431-4 ISBN 978-3-662-43432-1 (eBook)
DOI 10.1007/978-3-662-43432-1
Springer Heidelberg New York Dordrecht London

Library of Congress Control Number: 2014939621

Printed on acid-free paper

Springer is part of Springer Science+Business Media (www.springer.com)

Parts of this thesis have been published in the following journal articles:

- Wanai Li, Yu-Xin Ren, G. Lei and H. Luo, The multi-dimensional limiters for solving hyperbolic conservation laws on unstructured grids, J. Comput. Phys., 230(2011) 7775–7795
- Wanai Li, Yu-Xin Ren, The multi-dimensional limiters for solving hyperbolic conservation laws on unstructured grids II: Extension to high order finite volume schemes, J. Comput. Phys., 231(2012) 4053–4077
- Wanai Li, Yu-Xin Ren, High order *k*-exact WENO finite volume schemes for solving gas dynamic Euler equations on unstructured grids, Int. J. Numer. Meth. Fluids, 70(2012) 742–763
- Wanai Li, Yu-Xin Ren, The multi-dimensional limiters for discontinuous Galerkin method on unstructured grids, dx.doi.org/10.1016/j.compfluid.2013.12.008, accepted by Computer and Fluids, 2013
- Wanai Li, Yu-Xin Ren, Quadrature-free non-oscillation finite volume schemes for solving Navier–Stokes equations on unstructured grids, Proceeding of the Sixth International Conference on Fluid Mechanics, June 30–July 3, 2011, Guangzhou, China

Supervisor's Foreword

High-order numerical methods on unstructured grids, including the k-exact/WENO finite volume (FV), discontinuous Galerkin (DG), residual distribution, spectral volume/spectral difference, $P_N P_M$, and correction procedure via reconstruction methods, are widely considered as a major breakthrough in CFD. Intensive researches have been devoted in developing such kinds of numerical schemes. However, before some of these methods can be readily used as components of the next-generation CFD software, some bottleneck problems must be properly solved.

The goal of this thesis was to address some of these problems and propose simple yet effective solutions. One major focus is the development of shock capturing technique in terms of accuracy-preserving limiters. In order to be accuracy-preserving, several candidate reconstruction polynomials with the same order of accuracy should be provided for a given cell. In the WENO schemes, this is achieved by identifying several candidate stencils and performing reconstruction on each stencil. Because of the ambiguity in choosing multiple stencils and the cost of performing multiple reconstructions, the WENO approach on unstructured meshes is very complicated in coding and time-consuming in implementation. In this thesis, the secondary reconstruction (SR) procedure is proposed to provide the candidate polynomials used in the limiting procedure. The basic idea of the SR is to use the continuations of the existing solution polynomials (obtained from either the k-exact reconstruction in finite volume methods or the time integration of the weak formulations in the DG methods) on neighboring cells as the additional candidate reconstructions for the current cell. Therefore, there is no need to find multiple stencils. Furthermore, the polynomials of the SR can be derived analytically, which makes this approach much more efficient than the WENO schemes. To further improve the efficiency of the limiting procedure, the WBAP limiter is constructed to replace the weighted average procedure used in the WENO schemes. The WBAP limiter is inherently quadrature-free which makes it very efficient on the unstructured grids. Wanai Li studied the properties of WBAP limiter in detail and found two interesting limit formulations of the WBAP limiter. Using the SR and the WENO/WBAP procedures, very simple and robust limiters for the FV and DG methods have been designed. These limiters are compact, accuracy-preserving, and effective in removing numerical oscillations in the vicinities of the flow discontinuities.

Besides the limiters for shock capturing computation, high-order discretization of viscous terms, high-order algorithm for mixed elements, and high-order treatment of curved boundaries are also studied in this thesis and some very elegant solution procedures are proposed.

Beijing, January 2014 Yu-Xin Ren

Preface

When I come into contact with more people from industrial departments, the computation fluids dynamics (CFD) technique is found to be a standard tool in the engineering applications to my surprise. In the small-scale department such as air-conditioning/refrigerator manufactures, the CFD software begins to guide the design process. In the nuclear power plant, all kinds of CFD softwares are used to predict every possible accident. The aerospace department also searches for more reliable CFD methods to replace their expensive experiments. But how far should we trust these CFD data? The question is not easy to answer. The real flows are depicted using the Navier–Stokes equations which are strongly nonlinear. The flows contain large to very small scales which seem impossible to capture at the same time. Turbulence models are necessary to adopt to model these small-scale structures' behaviors. More advanced turbulent models such as large eddy simulation require high-order accurate numerical schemes. Some people predict that LES could be applied on engineering flows in 40 years with the current upgrade speed of computers. With high-order schemes, this period might be shorter. Besides, the configurations are always complex for engineering cases, thus unstructured grids are more flexible in the domain decompositions. Combining both these requirements, the high-order methods on unstructured grids seem a better choice to consider these problems. First of all, high-order schemes can capture the small-scale structures in high resolutions. Secondly, unstructured methods can work on complex geometries. Whereas these unstructured high-order methods are difficult to construct, this thesis tries to give some efficient implementations to make the high-order unstructured methods work robustly. I also believe that high-order accuracy is not only the trend for numerical methods, but also the requirement for other scientific investigation tools.

Guangzhou, China, December 2013 Wanai Li

Acknowledgments

This thesis was completed under the guidance of my supervisor Prof. Yu-Xin Ren and here I show him my great appreciation for all the help during my Ph.D. journey. These helps cover widely from the choosing of the research topic, guidance, good advice to conferences attending, etc. Not only have I gained knowledge from my tutor, but also the serious and strict attitude toward academic research. Now when I start teaching lessons in the university, my students are told about my experience in research attitude. Thanks should also be given to Prof. Hong Luo of North Carolina State University, Zhijian Wang of Kansas University for the kind discussion on the topic of unstructured grids methods.

For all the guys in our laboratory, I would say I'm so lucky to work with you. We lived together as brothers and sisters and shared so many happy moments.

Also, I need to thank my parents and brother for their love and encouragement during my growth. And to my wife I would say this thesis is also part of your work. Your support makes me feel warm even in the cold winter.

Lastly, many thanks are given to the projects 1172153 and 10932005 of NSFC for the financial support.

Contents

Chapter 1
Introduction

Over the last several decades, computational fluid dynamics (CFD) methods are undergoing significant developments and have developed to be one of the most important tools in the research of fluid dynamics in additional to theoretical analysis and experiments. The CFD technique finds wide usage in different fields related to fluids, e.g., aerodynamics force and heat flow, configuration optimization, flow control strategies, and turbulence strength enhancement. Among these applications, the AIAA holds five drag prediction workshops [1] and the latest High Lift workshop [2], which aim at calculating the lift and drag force accurately using these methods. To make the CFD methods reliable, a lot of research has been conducted on the validation and verification of CFD methods.

The second order finite volume/difference methods have now become mature and are widely used. Whereas with the increased requirement of engineering applications, second order schemes are far from adequate in predicting aerodynamics forces and vortex structures. High-order and high-resolution numerical schemes are being used to raise the computational efficiency. The ADIAGMA project at the sixth European research framework [3, 4] and the IDIHOM project [5] at the seventh European research framework propose to develop adaptive high-order variational methods for aerospace applications. The project Vision 2030 [6] also plans to construct the next generation of CFD software. The first high-order CFD workshop [7] set up all kinds of standard computational models to test the accuracy and computational abilities of high-order numerical methods in different problems. For simple cases, the high-order numerical schemes have already shown their great superiority. For e.g., the research by Dumbser [8] indicates that, when using high-order ADER-FV schemes in computing linear aeroacoustic problems, the accuracy of higher order schemes can be several orders higher than that of lower order schemes in the same computational time. For complex nonlinear equations such as Navier–Stokes equations, high-order schemes can obtain more accurate results in shorter times [9]. The research on high-order numerical methods has become a hotspot in CFD technique, which can be found widely in CFD journals and textbooks [10].

W. Li, *Efficient Implementation of High-Order Accurate Numerical Methods on Unstructured Grids*, Springer Theses, DOI: 10.1007/978-3-662-43432-1_1,

Some remarkable progress have been made in developing high-order numerical schemes, of which the finite difference method (FDM) was constructed first and applied in different complicated flows. To achieve similar resolutions as spectral methods, the compact FDM [11] has been proposed. Compact FDM uses fewer stencils than common FDM, but it is implicit in space discretization. Another scheme, DRP finite difference scheme [12], uses more stencils to achieve similar resolutions as compact FDM. When shock waves exist in the flow-field, high-order ENO/WENO [13] schemes are used to suppress the numerical oscillations near the shock waves and keep the high-order accuracy in the smooth region. However, ENO/WENO schemes bring certain numerical dissipations and degrade the resolutions of FDM. The existing way to deal with this problem is to use the high-resolution schemes in smooth flow region and the ENO/WENO schemes near the discontinuities, e.g., Ren's hybrid compact WENO scheme [14], Martin's bandwidth-optimized WENO scheme [15] and Sun's MDCD scheme [16]. These schemes have been applied on the direct numerical simulation of supersonic flows containing all kinds of flow structures.

However, the FDM only works on structured grids that are smoothly distributed. Thus the FDM has limited applications in treating flow problems with complex geometries. The structured grids can be generated by solving algebraic equations or differential equations. On complex geometries the multiblock or overset structured grids should be generated. The merit for structured grids is that the topology orthogonality can be controlled, while the demerits are that the grids need to be frequently adjusted manually as good and high-quality grids are difficult to generate on complex geometries. On the other hand, unstructured grids generation has overcome these defects. Commonly, it uses triangular or quadrilateral grids on two dimensions and tetrahedral or hexahedral grids on three dimensions. In fact, arbitrary polygons and polyhedra can be used as the control volume. Unstructured girds have a number of advantages. (1) They can be generated on very complex domains with high quality and fast speed. (2) They can be generated automatically. (3) It is easy to refine unstructured grids locally, which is beneficial for calculation of shock waves, shear layers, and the interface of two fluids. (4) It is natural to achieve a load-balanced parallelization on unstructured grids by using partitions of equal size.

Although the superiority of unstructured grids is obvious, we should realize that the construction of unstructured high-order methods is difficult. The reasons lie in many aspects. (1) Many efficient high-order schemes on structured grids cannot work well on unstructured grids. For e.g., the structured high-order FDM is constructed on one-dimensional equations and the analysis of FDM properties is relatively simple. Moreover, the extension of one-dimensional finite difference scheme to multi-dimensions can use the dimensional splitting technique. Because there is no obvious connection between the coordinate directions on unstructured grids, generalization to unstructured grids is extremely difficult. Another case is that the construction of shock capturing limiters is also based on one dimension, which is hard to extend to unstructured grids directly. (2) Unstructured high-order numerical methods couple with the complexity of geometries strongly, e.g., reconstruction for the physical data must consider the dimensions of grids, topologies, and geometrical characteristics.

With the increase in dimensions, the storage, CPU cost, and algorithm complexity for reconstruction also increase substantially.

To further analyze these issues, the following sections discuss the current development of unstructured high-order schemes and give a brief introduction to the main challenges.

1.1 High-Order Numerical Methods on Unstructured Grids

According to the construction methods for degree of freedom (DOF) of the polynomials inside the cell, the current unstructured numerical methods can be divided into the following catagories: methods using pure reconstruction, methods using inner DOFs, and methods using inner DOFs for reconstruction. The first one refers to the finite volume schemes. The second one includes the discontinuous Galerkin methods [17], spectral volume/difference method [18, 19] and Correction Procedure via Reconstruction (CPR) [20, 21], etc. And the third one refers to the $P_N P_M$ which is already applied in solving Navier–Stokes equations [8] and MHD equations [22]. A brief introduction is given below for these three methods to distinguish them from the ways in obtaining the DOFs.

Massive mature algorithms have been developed in the framework of second order finite volume methods (FVM), and the high-order FVM can also use these mature algorithms if the problems of high-order schemes are solved. Barth used the k-exact reconstruction to represent the distribution with high-order polynomials inside the 2D grids [23], which was generalized to 3D grids by Delanaye and Liu [24]. Abgrall extended the idea of ENO on structured grids to unstructured grids [25, 26], and in similar thought WENO was also extended to unstructured grids by Friedrich for better convergence and accuracy behavior [27]. The key component to achieve high-order accuracy using FVM is the high-order representation of physical data inside the cell. Because finite volume methods can only obtain the average inside the cell, the construction of high-order polynomials relies on the averages on the neighboring cells around the central one. To overcome the singularity of this construction, the number of construction stencils is much larger than DOFs as shown in Fig. 1.1a. The large stencils result in a series of problems in implementation: first, cache missing becomes serious due to the information in the stencils being far away in memory; second, large amount of data need to be transferred in the interface of two partitions, which would lead to bad efficiency in parallel computation. Moreover, to reduce the computation, the general inverse matrix for solving the k-exact reconstruction using least-square method requires to be stored, which takes up too much memory.

The appearance of spectral volume method (SVM) is mainly to address the problem of large stencils. It can be viewed as a generalization of FVM to high-order schemes, so the mature algorithms on finite volume methods can also be applied on spectral volume methods. As shown in Fig. 1.1b, being different from FVM which chooses the stencils outwards, the SVM constructs the polynomials through dividing subcells inside the control volume. The main work of SVM is designing of reasonable

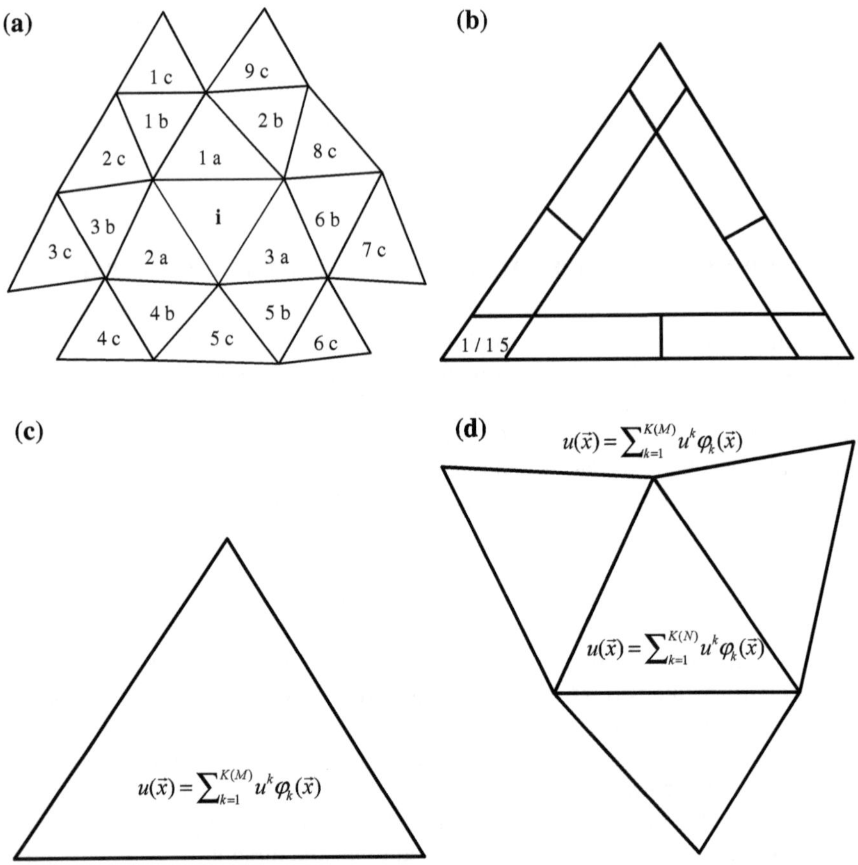

Fig. 1.1 The stencils for different methods using fourth order scheme. **a** Finite volume method **b** Spectral volume method **c** Discontinuous Galerkin method **d** $P_N P_M$

division of subcells which would make the scheme nonsingular and achieve better stability. Whereas the subcell division for high-order schemes in two or three dimensions is not easy. Strict analysis [28] proves that, on two-dimensional triangular cells, the second order SVM can be stable yet the third and fourth order SVMs are weakly unstable. Spectral difference method (SDM) uses the solution point distributions instead of subcells, which simplifies the construction of high-order schemes. Moreover, SDM can adopt a similar way of calculation for flux divergence in finite difference method, which is easily implemented and computationally efficient. Nevertheless, a strict analysis [29] shows that, for two-dimensional triangular grids, second order SDM can be stable by choosing proper solution point distributions and flux computation way, yet third order and fourth order SDMs are unstable. For quadrilateral grids, the SDM is stable using the tensor product of one-dimensional spectral difference schemes. To achieve better stability on triangles using SDM, the latest work would be the new flux interpolation method in terms of Raviart–Thomas space [30].

The discontinuous Galerkin method directly gives the the high-order polynomials approximations inside the cell, and computes the coefficients of the polynomials using the weak form of the governing equations. As shown in Fig. 1.1c, the construction uses the inner DOF u_l to represent the polynomial coefficient. Similar to SVM/SDM, DGM overcomes the trouble of choosing stencils in FVM. Furthermore, DGM is proved to possess the mathematical properties with respect to conservation, stability, small dissipation, and superconvergence for some problems [31]. These properties make DGM the most popular unstructured high-order methods nowadays. While comparing with FVM and SVM, DGM requires higher order quadrature rules for face and volume integration and smaller time steps in the time evolution. So the urgent need is to reduce the computational cost in DGM. Among the improvement techniques, one is to use the nodal discontinuous Galerkin method, which stores the physical quantities on the quadrature points of the volume integration and thus saves the cost in evaluating the data on these quadrature points. The Correction Procedure via Reconstruction method proposed by Huynh [21] and Wang [20] can be viewed as a general type of nodal discontinuous Galerkin method using strong conservation form.

The $P_N P_M$ method proposed by Dumbser is based on the modal discontinuous Galerkin method. The low-order DOFs of $P_N P_M$ adopt the ones of DGM, while the high-order DOFs are reconstructed using the lower order DOFs from face-neighbor cells. Also, $P_N P_M$ combines the FVM and DGM in a uniform framework, i.e., $P_0 P_M$ represent the $(M+1)$-th order FVM and $P_M P_M$ represent the $(M+1)$-th order DGM. When N and M are chosen properly, the $P_N P_M$ method not only avoids the large stencils problem, but also reduces the number of equations for solving DOFs and enlarges the time step in DGM. To make the stencils compact, the $P_N P_M$ method only chooses the face-neighbor cells as shown in Fig. 1.1d.

1.2 Applications and Challenges for Unstructured High-Order Methods

Unstructured high-order methods are in demand in real engineering applications due to the flexibility of unstructured grids. The ADIGMA project [3] in Europe aims at developing reliable CFD numerical methods to simulate the real flows. One key point of this project is to obtain the grid-independent solutions. Developing high-order numerical schemes is the most efficient way to achieve this target. So the ADIGMA project proposes to use the popular DGM and Continuous Residual Distribution (CRD) in solving engineering problems. In the final report of the ADIGMA project [4], the high-order methods are applied in computing the transonic viscous flows around the aircraft by solving the Reynolds averaged Navier–Stokes equations. But the problems of convergence and shock capturing are still waiting for improvement.

Collis [32] and Wei [33] employ the DGM in the direction numerical simulations of channel flow. Zhou and Wang [34] apply the SDM in large eddy simulations of SD7003 airfoil. The aeroacoustic problem requires low dissipation and low dispersion schemes to capture the pressure fluctuations. And in complex configurations, unstructured grids are always needed to discretize the flow regions, e.g., the references [35, 36] use the FV-MLS and SUPG methods for solving the aeroacoustic problems on unstructured grids, respectively.

Although unstructured high-order methods are used to get better computational efficiency, they are not widely used in engineering applications. The overall reasons for this phenomenon are that the high-order schemes are not as robust as lower order ones and the implementation process are much more complicated, e.g., for the steady-state problem, the convergence speed using high-order methods would be slower or even diverge due to the rigidness of high-order discretized equations. In flows with strong shock waves and rarefaction, mature techniques are developed for low-order schemes to suppress the oscillations near the shock waves and assure positivity. Nevertheless, similar techniques for high-order schemes are far from mature. To make high-order schemes serve for real engineering applications, the following work still needs further improvement.

1.2.1 Computational Cost and Storage

The computational cost and storage for high-order schemes are greatly increased compared to low-order schemes on the same grids. The reasons can be summarized as follows. (1) The ways for obtaining the high-order polynomials distributions inside the cell are complicated. For FVM, the complexity lies in the large stencils in the reconstruction process and the overdetermined linear equations solved using least-square methods. The enlargement of stencils would lead to cache missing problem and the linear equations solving needs large computation and storage, e.g., the WENO schemes on unstructured grids need the identification of one central stencil and several biased stencils (the WENO schemes improved by Dumbser [37] use seven biased stencils to construct the polynomials). (2) The number of Gaussian points for the flux computation would increase, i.e., the Riemann solver or flux splitting will increase. (3) The weak form using DGM introduces the source terms which requires the high-order volume integration. Usually, the DOFs are collocated on the quadrature points and quadrature-free implementations are developed for less computational cost.

1.2.2 Boundary Treatment

The difficulties come from two aspects. One is that for the reconstruction methods, e.g., FVM and $P_N P_M$ method, the singularity of reconstruction on the boundary becomes serious due to the biased stencils. And the solution is added more stencils into reconstruction to make the scheme stable. The special treatment on the boundary is not in accord with the inner cells and makes the implementation difficult. What is more serious is that we cannot guarantee the stability of the scheme.

The other difficulty is that the real curved configurations are required to be taken into account for the curved wall boundary. Bassi and Rebay [38] used the DGM to solve the two-dimensional inviscid flow around the curved wall surface, and discovered that the wall approximation must also achieve high-order accuracy. In their calculations, straight wall approximation would lead to nonphysically results. Similarly, straight wall approximation in three-dimensional flow computations would also give completely wrong results [39]. Other high-order schemes like FVM also need similar requirements [40]. The curved boundary treatment needs to know the actual wall configuration. This can be attained by outputting more inner points on the wall face or reconstructing the curved wall using the straight wall face and faces nearby, both of which are difficult in implementation for real complex geometries. When the high-order approximation is done, the high-order schemes need quadrature on the curved wall face or cell with curved wall. For the simplex, isoparametric transformation is needed for the quadrature rule. Whereas the quadrature also becomes complicated for the complex control volumes.

To overcome this difficulty, Krivodonova et al. [41] put forwards a novel treatment way. In their method, the control volume still chooses the one with straight face, but the normal vectors in computing the flux on the curved surface approximate the real ones with high-order accuracy. This approximation is easy to implement for the flux on the wall surface and inner surface is nearly the same. Also the results on two-dimensional inviscid flows around curved wall surface achieve high-order accuracy. The high-order approximation on 2D grids is easy, but it becomes extremely difficult on 3D grids. Luo [42] reads the geometry information while generating the grids thus gives the real normal vectors. Nevertheless, reading the geometry information is also a tough job. And currently few references use this method to treat the viscous flows, especially with nonslip, adiabatic/isothermal wall boundary conditions.

1.2.3 Shock Capturing Technique

No matter the high-order reconstruction of FVM or the polynomials using DOFs in DGM, are all based on such a premise that the distribution of physical data is continuous on the stencils. Whereas there exist many types of discontinuities in the compressible flows such as shock waves, contact discontinuities, and head and rear of rarefaction. The discontinuous distribution would generate numerical oscillations in the vicinities of discontinuities and pollute the accuracy on other smooth regions. These oscillations would even lead to negative pressures or densities and the computation diverges. When the order of the scheme gets higher, the oscillations become severe. So a major challenge is to keep the monotonicity of the scheme near the discontinuities. Moreover, it is also a main technique to solve the engineering cases robustly.

There are two main techniques to suppress the oscillations near the discontinuities. One is the artificial viscosities to add more dissipation near the discontinuities, and the other is based on the reconstruction and limiting technique. The merit of artificial viscosities is that it can make the scheme compact for it only needs the information on local cells such as pressure gradients, while the defect is that the amount of artificial viscosity needs to adjust according to the problems. Over-added artificial viscosities would sacrifice the accuracy of the solution and insufficient viscosity cannot suppress the numerical oscillations. Besides, the adding of artificial viscosities would decrease the time step and make the convergence slow.

The reconstruction and limiting technique is some kind of postprocess. It directly modifies the polynomials in the cell such as:

$$\tilde{u}(\mathbf{x}) = \phi u(\mathbf{x})$$

where $\phi \in [0, 1]$. The high-order limiter is mature for the finite difference/volume methods on structured grids. The ENO/WENO FDM and FVM [13, 43] on structured grids can suppress the oscillations in high resolution, and keep the high-order accuracy in smooth regions in the meantime. On unstructured grids, the reconstruction and limiting technique for the second order FVM is also widely used such as the Barth limiter [44], Venkatanishnan limiter [45] and MLP limiter [46]. Whereas the high-order limiters on unstructured grids are extremely difficult, e.g., the finite volume WENO limiter [25, 27, 47] on unstructured grids requires the reconstruction using biased stencils in every direction which is ambiguous and computationally expensive. The limiter for methods using inner DOFs such as DGM and SVM/SDM, use the limiters of FVM in the following ways. The first is to use shock detector in detecting the trouble cells, and apply the second order FV limiters on the shock regions(Luo [42] and Wang [18]). This way keeps the compact property but would damage the accuracy in smooth region if marked as trouble cells. The other is to develop high-order limiting technique such as WENO/Hermite WENO [48] for discontinuous Galerkin method. The accuracy of this way can improve comparing to second order limiter, but it destroys the compactness of the original methods. The high-order compact limiters, e.g., moment limiter [49], hierarchical & pointwise hierarchical [50, 51], reconstruction and DG-MLP [52] limiter, is still under development.

The high-order limiting always requires applying on characteristic variables for the nonlinear systems which makes the limiting much more expensive. Since the discontinuities are the local phenomenon in the regions, a shock detector is usually used to distinguish the trouble cells and the limiting is only applied on the trouble cells. The frequently used shock detectors are the TVB shock detectors [17], BSB [53] and KXRCF [54] detectors.

1.2.4 Efficient Time Integration Strategy

The time integration can be explicit or implicit, both of which have their defects for high-order space discretization. The explicit scheme can achieve high-order time accuracy, e.g., the RKDG developed by Cockburn and Shu uses the multistep Runge-Kutta method [17], the ADER-FV and ADER-DG schemes developed by Toro and Dumbser combine the time discretization and space discretization and achieve arbitrary high-order accuracy [8, 55]. The problem of the explicit scheme is that the time step is restricted by the CFL condition and global time step is needed for unsteady problem. These restrictions would make the simulations for the unsteady boundary layer flows too slow. The time step for implicit time integration is not restricted by CFL condition. In fact, the time step can usually be infinitely large theoretically. Nevertheless, the solving for the implicit methods often requires Newton type iteration methods to solve large-scale algebraic equations, which is computationally expensive and requires large amount of storage. The current develop trend is to combine the explicit and implicit time integration such as the implicit–explicit Runge-Kutta methods [56].

1.3 The Target of This Thesis

To develop the practical high-order unstructured numerical schemes for engineering applications, the above challenging issues must need to be addressed first. This thesis mainly studies the construction of the shock capturing limiters and the high-order treatment of curved wall boundary. The reason for choosing the two issues is that they are the common bottlenecks for different kinds of high-order numerical methods and affect the accuracy, efficiency, and robustness deeply. In addition, this thesis also involves other issues related to the practical applications for the unstructured high-order methods such as the mixed grids, viscous terms calculations. The high-order methods studied in this thesis are FVM and DGM, which can be viewed as two extreme cases of high-order methods. FVM only relies on the reconstruction to give the polynomial distributions, while DGM uses the weak form of governing equations to evolve the DOFs of polynomials. In the meantime, the applications of these two methods are more widely studied than other types of high-order methods. So the study of these two methods makes the applications more general.

When discontinuities exist in the flow-field, the high-order schemes would generate large numerical oscillations in the vicinities of discontinuities. The current shock capturing schemes would degrade the accuracy in smooth regions, or destroy the nice properties such as compactness in DGM, or lead to obvious increase in CPU and memory cost such as WENO schemes. This thesis tries to develop new types of limiter and limiting process to solve the above problems. The WENO scheme is one of the best shocks capturing technique on unstructured grids presently. Its main characteristic is the multiple reconstructions and weighted average for the candidate polynomials. These properties make it possesses the fine shock capturing abilities but are also its main defect, i.e., the cause of complicated implementation, large CPU, and memory cost. This thesis will propose a new way of constructing the candidate polynomials, which would be easily and efficiently implemented. Meanwhile this thesis focuses on developing a novel compact and quadrature-free weighted procedure. Based on systematic and theoretical analysis, the thesis will create a simple, compact and accuracy preserving limiting strategy, and apply it on the unstructured high-order FVM and DGM.

The geometries in engineering problems are always rather complicated, so the curved wall boundaries must be treated well in the high-order schemes. If the real curved wall shape cannot be considered in the computation, the accuracy and robustness will be degraded, or even nonphysical solutions come out in the results. There are two main problems in treating the curved boundary, namely the accurate representation for the curved surface and the high-order construction of curved cells near the boundaries. The former needs the update of the grids generation software, while the latter will have more theoretical values. This thesis devotes to the latter. The specific strategy used in this thesis is to use isoparametric transformation to change the curved cells to the regular cells, and then the quadrature on the curved cells can be evaluated. The traditional transformation requires too much information which brings expensive computational cost. This thesis tries to give some simplified transformations which need less control points and coincide with the inner cells with straight sides.

In order to construct a complete and efficient high-order numerical scheme, we still need to supply other parts of the numerical framework. The traditional viscous terms computation is rather complicated and time consuming in FVM. This thesis utilizes the polynomial reconstruction in high-order FVM and proposes an efficient way for computing the viscous flux with high-order accuracy. In the viscous flows, usually grids with large aspect ratio in flow and normal directions are needed to assure the computational accuracy. This is achieved using mixed elements in the unstructured grids computations, namely triangular/quadrilateral grids on two dimensions, tetrahedral/prismatic/hexahedral grids on three dimensions. This thesis also studies the implement on mixed grids. In fact, some algorithms is hard to or cannot be realized on mixed elements, e.g., the WENO schemes of Hu and Shu [47] is only applied on triangular cells, the quadrature-free WENO scheme of Dumbser [57] works only on triangular or tetrahedral grids. So the application and efficiency of the newly proposed schemes on mixed elements are also main studied problems in this thesis. Through the efforts in addressing all these issues, the efficiency of the high-order numerical schemes can be raised and the schemes are capable to work on the simulations of complicated flow problems.

References

1. Levy DW, Laflin KR, Tinoco EN, et al (2013) Summary of data from the fifth AIAA CFD drag prediction workshop. AIAA Paper 2013-0046, 51st AIAA Aerospace Sciences Meeting, 7–10 Jan 2013. Grapevine, TX, United States
2. Slotnick JP, Hannon JA, Chaffin M (2011) Overview of the first AIAA CFD high lift prediction workshop (Invited). AIAA 2011-862, 49th AIAA Aerospace Sciences Meeting including the New Horizons Forum and Aerospace Exposition, 4–7Jan 2011. Orlando, Florida
3. Kroll N (2006) ADIGMA—a European project on the development of adaptive higher order variational methods for aerospace applations. European Conference on Computational Fluid Dynamics, ECCOMAS CFD 2006
4. Kroll N (2009) ADIGMA—a European project on the development of adaptive higher-order variational methods for aerospace applications, 47th AIAA Aerospace Sciences Meeting. 2009-176
5. Industrialisation of High-Order Methods A Top-Down Approach (IDIHOM), FP7 reference number:265780, Duration: 2010.10.1 - 2013.09.30
6. Research Opportunities in Aeronautics (ROA) (2011) NASA Research Announcement (NRA):NNH10ZEA001N, August 26, 2011
7. Wang ZJ, Fidkowski K, Abgrall R, et al (2012) High-order CFD methods: current status and perspective. Int J Numer Meth Fluids: 1–42
8. Dumbser M, Schwartzkop T, Munz CD (2006) Arbitrary high order finite volume scheme for linear wave propagation. Book series notes on numerical fluid mechanics and multidisciplinary design, vol 91, pp 129–144
9. Dumbser M (2011) Arbitrary high order PNPM schemes on unstructured meshes for the compressible Navier-Stokes equations. Comput Fluids 39:60–76
10. Wang ZJ (2011) Adaptive high-order methods in computational fluid dynamics. World Scientific, New Jersey
11. Lele SK (1992) Compact finite difference scheme with spectral like resolution. J Comput Phys 103:16–42
12. Tam C, Webb JC (1993) Dispersion-relation-preserving finite difference schemes for computational acoustics. J Comput Phys 107:262–281
13. Shu CW Essentially non-oscillatory and weighted essentially non-oscillatory schemes for hyperbolic conservation laws. NASA/CR-97-206253, ICASE, Report No. 97–65
14. Ren YX, Liu M, Zhang HX (2003) A characteristic-wise hybrid compact-WENO scheme for solving hyperbolic conservation laws. J Comput Phys 192:365–386
15. Martin MP, Taylor EM, Wu M et al (2006) A bandwidth-optimized WENO scheme for the effective direct numerical simulation of compressible turbulence. J Comput Phys 220:270–289
16. Sun ZS, Ren YX, Larricq C et al (2011) A class of finite difference schemes with low dispersion and controllable dissipation for DNS of compressible turbulence. J Comp Phys 230:4616–4635
17. Cockburn B, Shu CW (2001) Runge-Kutta discontinuous Galerkin methods for convection-dominated problems. J Sci Comput 16:173–261
18. Wang ZJ, Zhang L, Liu Y (2004) Spectral (finite) volume method for conservation laws on unstructured grids IV: extension to two-dimensional systems. J Comput Phys 194:716–741
19. Liu Y, Vinokur M, Wang ZJ (2006) Spectral difference method for unstructured grids I: basis formulation. J Comput Phys 216:780–801
20. Wang ZJ, Gao H (2009) A unifying lifting collocation penalty formulation including the discontinuous Galerkin, spectral volume/difference methods for conservation laws on mixed grids. J Comput Phys 228:8161–8186
21. Huynh HT (2007) A flux reconstruction approach to high-order schemes including discontinuous Galerkin methods. 18th AIAA Computational Fluid Dynamics Conference, 25–28 June 2007. Miami, FL, AIAA2007-4079
22. Dumbser M, Zanotti O (2009) Very high order PNPM schemes on unstructured meshes for the resistive relativistic MHD equations. J Comput Phys 228:6991–7006

23. Barth TJ, Frederichson PO (1990) Higher order solution of the Euler equations on unstructured grids using quadratic reconstruction. 28th aerospace sciences meeting, 8–11 Jan. Reno, Nevada, AIAA90-0013
24. Delanaye M, Liu Y (1999) Quadratic reconstruction finite volume schemes on 3D arbitrary unstructured polyhedral grids. 14th Computational Fluid Dynamics Conference, AIAA99-3259
25. Abgrall R (1994) On essentially non-oscillatory schemes on unstructured meshes: analysis and implementation. J Comput Phys 144:45–58
26. Abgrall R, Sonar T (1997) On the use of Mühlbach expansions in the recovery step of ENO methods. Numer Math 76:1–25
27. Friedrich O (1998) Weighted essentially non-oscillatory schemes for the interpolation of mean values on unstructured grids. J Comput Phys 144:194–212
28. Van Abeele K, Lacor C (2007) An accuracy and stability study of the 2D spectral volume method. J Comput Phys 226:1007–1026
29. Van Abeele KV, Lacor C, Wang ZJ (2008) On the stability and accuracy of the spectral difference method. J Sci Comput 37:162–188
30. Balan A, May G, Schöberl J (2011) A stable spectral difference method for triangles. 49th AIAA aerospace sciences meeting including the new horizons forum and aerospace exposition, 4–7 Jan. Orando, Florida, AIAA2011-47
31. Cockburn B, Kanschat G, Perugia I et al (2011) Superconvergence of the local discontinuous Galerkin method for elliptic problems on Cartesian grids. SIAM J Numer Anal 39:264–285
32. Collis SS, Ghayour K (2003) Discontinuous Galerkin methods for compressible DNS. In: Proceedings of FEDSM, ASME/JSME joint fluids engineering conference, 2003
33. Wei L, Pollard A (2011) Direct numerical simulation of compressible turbulent channel flows using the discontinuous Galerkin method. Comput Fluids 44:85–100
34. Zhou Y, Wang ZJ (2010) Implicit large eddy simulation of transitional flow over a SD7003 wing using high-order spectral difference method. 40th fluid dynamics conference and exhibit, 28 June–1 July. Chicago, Illinois, AIAA2010-4442
35. Nogueira X et al (2010) Resolution of computational aeroacoustics problems on unstructured grids with a higher-order finite volume scheme. J Comput App Math 234:2089–2097
36. Rao PP (2004) High order unstructured grid methods for computational aeroacoustics. Phd thesis, United States: the Pennsylvania State university
37. Dumbser M, Käser M (2007) Arbitrary high order non-oscillatory finite volume schemes on unstructured meshes for linear hyperbolic systems. J Comput Phys 221:693–723
38. Bassi F, Rebay S (1997) High-order accurate discontinuous finite element solution of the 2D Euler equations. J Comput Phys 138:251–285
39. Lübon C, Wagner S (2007) Three-dimensional discontinuous Galerkin codes to simulate viscous flow by spatial discretization of high order and curved elements on unstructured grids. New Res In Num Exp Fluid Mech 96:145–153
40. Ollivier-Gooch C, Altena MV (2002) A high-order-accurate unstructured mesh finite-volume scheme for the advection-diffusion equation. J Comput Phys 181:729–752
41. Krivodonova L, Berger M (2006) High-order accurate implementation of solid wall boundary conditions in curved geometries. J Comput Phys 211:492–512
42. Luo H, Baum JD, Löhner R (2008) On the computation of steady-state compressible flows using a discontinuous Galerkin method. Int J Numer Meth Eng 73:597–623
43. Jiang GS, Shu CW (1996) Efficient implementation of weighted ENO schemes. J Comput Phys 126:202–228
44. Barth TJ, Jespersen D (1989) The design and application of upwind schemes on unstructured meshes. In: Proceedings of the 27th AIAA aerospace sciences meeting, Reno, NV, Paper AIAA 89–0366, 1989
45. Venkatakrishnan V (1995) Convergence to steady state solutions of the Euler equations on unstructured grids with limiters. J Comput Phys 118:120–130
46. Park JS, Yoon SH, Kim C (2010) Multi-dimensional limiting process for hyperbolic conservation laws on unstructured grids. J Comput Phys 229:788–812

47. Hu C, Shu CW (1999) Weighted essentially non-oscillatory schemes on triangular meshes. J Comput Phys 150:97–127
48. Zhu J, Qiu J (2009) Hermite WENO schemes and their application as limiters for Runge-Kutta discontinuous Galerkin method III: unstructured meshes. J Sci Comput 39:293–321
49. Krivodonova L (2007) Limiters for high-order discontinuous Galerkin methods. J Comput Phys 226:879–896
50. Xu Z, Liu Y, Shu CW (2009) Hierarchical reconstructin for discontinuous Galerkin methods on unstructured grids with a WENO-type linear reconstruction and partial neighboring cells. J Comput Phys 228:2194–2212
51. Xu Z, Liu Y, Shu CW (2011) Point-wise hierarchical reconstruction for discontinuous Galerkin and finite volume methods for solving conservation laws. J Comput Phys 230:6843–6865
52. Park JS, Kim C (2011) Higher-order discontinuous Galerkin-MLP methods on triangular and tetrahedral grids. In: Proceedings of AIAA 2011–3059, 20th AIAA computational fluid dynamics conference, 27–30 June 2011, Honolulu, Hawaii
53. Burbeau A, Sagaut P, Bruneau CH (2011) A problem-independent limiter for high-order RungeCKutta discontinuous Galerkin methods. J Comput Phys 169:111C150
54. Krivodonova L et al (2004) Shock detection and limiting with discontinuous Galerkin methods for hyperbolic conservation laws. Appl Num Math 48:323–338
55. Dumbser M, Käser M, Toro EF (2007) An arbitrary high order discontinuous Galerkin method for elastic waves on unstructured meshes V: local time stepping and p-adaptivity. Geophys J Int 171:695C717
56. Kanevsky A et al (2007) Application of implicit-explicit high-order Runge-Kutta methods to discontinuous Galerkin schemes. J Comput Phys 225:1753–1781
57. Dumbser M, Käser M, Titarev VA et al (2007) Quadrature-free non-oscillatory finite volume schemes on unstructured meshes for nonlinear hyperbolic systems. J Comput Phys 226:204–243

Chapter 2
High-Order Finite Volume Method for the Compressible Flows

This chapter will introduce the basis framework of FVM and the governing equations are the compressible Navier-Stokes equations. The k-exact reconstruction will be specifically described since it is the key for FVM to achieve high-order accuracy. To solve the viscous flows, a new simple yet an efficient scheme is proposed and the viscous boundary conditions are considered through the constrained least-square method. For faster convergence speed, the dual time stepping with LU-SGS iteration is used for time integration. Also MPI parallelization is implemented in the program for solving large-scale problems. For brevity, only the three-dimensional description for the algorithms below is given below. Those two-dimensional descriptions being distinct from three-dimensional ones are also presented.

2.1 The Framework of the Finite Volume Method

This section will describe the basis framework of cell-center finite volume method on unstructured grids. The governing equations of compressible flows are the Navier-Stokes equations,

$$\frac{\partial U}{\partial t}+\frac{\partial\left(F-F_v\right)}{\partial x}+\frac{\partial\left(G-G_v\right)}{\partial y}+\frac{\partial\left(H-H_v\right)}{\partial z}=0 \tag{2.1}$$

where the conservative variables and inviscid flux vectors are

$$U=\begin{pmatrix}\rho\\ \rho u\\ \rho v\\ \rho w\\ E\end{pmatrix},\quad F=\begin{pmatrix}\rho u\\ \rho u^2+p\\ \rho uv\\ \rho uw\\ u(E+p)\end{pmatrix},\quad G=\begin{pmatrix}\rho v\\ \rho uv\\ \rho v^2+p\\ \rho vw\\ v(E+p)\end{pmatrix},\quad H=\begin{pmatrix}\rho w\\ \rho uw\\ \rho vw\\ \rho w^2+p\\ w(E+p)\end{pmatrix}, \tag{2.2}$$

and the viscous flux vectors are

W. Li, *Efficient Implementation of High-Order Accurate Numerical Methods on Unstructured Grids*, Springer Theses, DOI: 10.1007/978-3-662-43432-1_2,

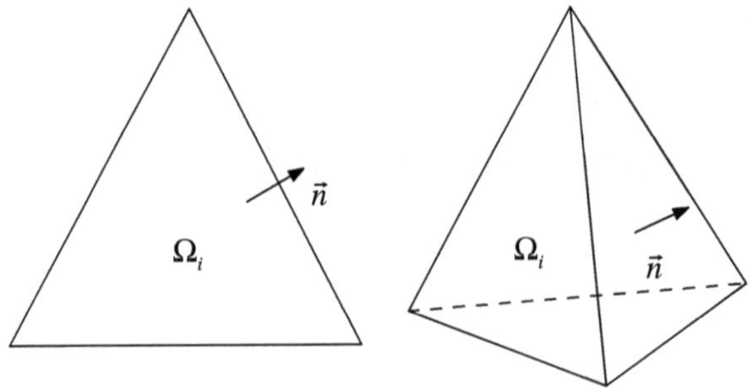

Fig. 2.1 The control volumes on two and three dimensions and the normal vector outwards

$$
\begin{aligned}
F_v &= \left\{0,\, \tau_{xx},\, \tau_{yx},\, \tau_{zx},\, u\tau_{xx} + v\tau_{yx} + w\tau_{zx} + k\frac{\partial T}{\partial x}\right\}^T, \\
G_v &= \left\{0,\, \tau_{xy},\, \tau_{yy},\, \tau_{zy},\, u\tau_{xy} + v\tau_{yy} + w\tau_{zy} + k\frac{\partial T}{\partial y}\right\}^T, \\
H_v &= \left\{0,\, \tau_{xz},\, \tau_{yz},\, \tau_{zz},\, u\tau_{xz} + v\tau_{yz} + w\tau_{zz} + k\frac{\partial T}{\partial z}\right\}^T.
\end{aligned}
\tag{2.3}
$$

with the viscous stress

$$
\begin{aligned}
\tau_{xx} &= 2\mu u_x + \lambda(u_x + v_y + w_z), \qquad \tau_{xy} = \tau_{yx} = \mu(u_y + v_x), \\
\tau_{yy} &= 2\mu v_y + \lambda(u_x + v_y + w_z), \qquad \tau_{xz} = \tau_{zx} = \mu(u_z + w_x), \\
\tau_{zz} &= 2\mu w_z + \lambda(u_x + v_y + w_z), \qquad \tau_{yz} = \tau_{zy} = \mu(v_z + w_y).
\end{aligned}
$$

In the above equations, ρ denotes the density, (u, v, w) are the velocity vector, μ is the dynamics viscosity, T is the temperature, E is the total energy and p is the pressure satisfying

$$
E = \frac{p}{\gamma - 1} + \frac{1}{2}\rho\left(u^2 + v^2 + w^2\right)
$$

with a constant ratio of specific heat $\gamma = 1.4$. According to the Stokes hypothesis, one has $\lambda = -2/3$. The heat conductivity is computed using $k = c_p\mu/Pr$, where c_p is the specific heat at constant pressure and Pr is Prandtl number.

For the control volume shown in Fig. 2.1, integrate over the control volume and get

$$
\frac{\mathrm{d}}{\mathrm{d}t}\bar{U}_i + \frac{1}{\bar{\Omega}_i}\oint_{\partial\Omega_i} \mathbf{F}_n \cdot \mathbf{n}\,\mathrm{d}S = 0, \quad \mathbf{F}_n = (F - F_v, G - G_v, H - H_v) \tag{2.4}
$$

where $\bar{\Omega}_i$ is the volume of cell i, $\mathbf{n}$ is the outward unit vector normal to the surface $\partial\Omega_i$, $\bar{U}_i$ is the average for conservative variables on the control volume defined as

Table 2.1 Number of Gaussian points on the triangle and tetrahedron

Order	Triangle			Quadrilateral		
	$K = 1$	$K = 2$	$K = 3$	$K = 1$	$K = 2$	$K = 3$
Surface integral	1	2	2	1	3	4
Volume integral	1	3	4	1	4	5

$$\bar{U}_i = \frac{1}{\bar{\Omega}_i} \int_{\Omega_i} U(x, y, z)\, d\Omega$$

The surface integral on can be expressed as the summation of integrals on all faces, i.e.,

$$\oint_{\partial\Omega_i} \mathbf{F}_n \cdot \mathbf{n}\, dS = \sum_{k=1}^{N_F} \int_{\partial\Omega_{i,k}} \mathbf{F}_n \cdot \mathbf{n}\, dS \approx \sum_{k=1}^{N_F} \left(\sum_{\phi=1}^{N_G} \omega_g \hat{F}_n(U^R(\mathbf{x}_{k,\phi}), U^L(\mathbf{x}_{k,\phi})) \cdot \mathbf{n} S_k \right) \tag{2.5}$$

where the left and right states, U^R, U^L, on the face can be obtained using the high-order k-exact reconstruction in Sect. 2.2, N_F is the number of faces for cell i. The flux integral is computed using the Gaussian integration and N_G is the number of Gaussian integration points. And the number of integration points for the surface and volume integration on different shapes of control volume is shown in Table 2.1. The numerical flux $\hat{F}$ is evaluated using the Riemann solver, for which the Roe and HLL schemes are used in this thesis. The calculation for the viscous flux will be introduced in Sect. 2.3.

2.2 The Time Integration Schemes

For the semi-discretization schemes of (2.4), the time integration scheme is needed to update the cell average. In this chapter, the explicit time integration is first considered. Specifically, the third- and fourth-order Runge-Kutta methods are chosen to agree with the order of the spatial discretization. For the explicit time integration, the time step of FVM is

$$\Delta t = CFL \frac{\bar{\Omega}_i}{\sum_{f \in \partial\Omega_i} \lambda_{c,f} S_f + 2\lambda_{v,f} S_f^2 / \bar{\Omega}_i}, \quad CFL \leq 1$$

where $\lambda_c = |\mathbf{v} \cdot \mathbf{n}_f| + c$ is the spectral radius of the flux Jacobian matrix at the cell face, c is the speed of sound and $c^2 = \gamma RT$, the diffusion coefficient is $\lambda_v = \max(\frac{4}{3}, \frac{\gamma}{\Pr})\frac{\mu}{\rho}$. The finite volume scheme (2.4) can be rewritten as the following semi-discretization form

$$\bar{\Omega}_i \frac{\mathrm{d}\bar{U}_i}{\mathrm{d}t} = R(U) \tag{2.6}$$

where the term on the right-hand side is

$$R(U) = -\frac{1}{\bar{\Omega}_i} \oint_{\partial\Omega_i} \mathbf{F}_n \cdot \mathbf{n}\,\mathrm{d}S$$

Then the third order explicit TVD Runge-Kutta methods for (2.6) can be written as

$$\begin{aligned}
U^{(0)} &= \bar{U}_i^n, \\
U^{(1)} &= U^{(0)} + \frac{\Delta t}{\bar{\Omega}_i} \cdot R(U^{(0)}), \\
U^{(2)} &= \frac{3}{4} U^{(0)} + \frac{1}{4}\left(U^{(1)} + \frac{\Delta t}{\bar{\Omega}_i} \cdot R(U^{(1)})\right), \\
U^{(3)} &= \frac{1}{3} U^{(0)} + \frac{2}{3}\left(U^{(2)} + \frac{\Delta t}{\bar{\Omega}_i} \cdot R(U^{(2)})\right), \\
\bar{U}_i^{n+1} &= U^{(3)}
\end{aligned} \tag{2.7}$$

and the fourth order Runge-Kutta method as

$$\begin{aligned}
U^{(0)} &= \bar{U}_i^n, \\
U^{(1)} &= U^{(0)} + \frac{\Delta t}{2\bar{\Omega}_i} R\left(U^{(0)}\right), \\
U^{(2)} &= U^{(0)} + \frac{\Delta t}{2\bar{\Omega}_i} R\left(U^{(1)}\right), \\
U^{(3)} &= U^{(0)} + \frac{\Delta t}{\bar{\Omega}_i} R\left(U^{(2)}\right), \\
U^{(4)} &= U^{(0)} + \frac{\Delta t}{6\bar{\Omega}_i}\left(R\left(U^{(0)}\right) + 2R\left(U^{(1)}\right) + 2R\left(U^{(2)}\right) + R\left(U^{(3)}\right)\right), \\
\bar{U}_i^{n+1} &= U^{(4)}.
\end{aligned} \tag{2.8}$$

The defect of the explicit time integration is that it is restricted by the CFL condition, which makes the time step small. In simulating the engineering flows, the grids are always refined near the regions of shock wave, boundary layers and strong vortex, etc. In these situations, the global time step is very small and the explicit time integration is not a good choice in efficiency. On the other hand, the implicit time integration can break the limit of CFL condition. The much larger time step, theoretically speaking, infinitely large for fully implicit scheme, can speed up the computation greatly. The following section will show how to construct the LU-SGS implicit dual time-stepping algorithm for the Navier-Stokes equations [1, 2].

Use the semi-discretization scheme in (2.6) and have

$$\frac{\partial \bar{U}_i^n}{\partial t} = R(U_i^{n+1})/\bar{\Omega}_i$$

Introduce the term of virtual time $\partial \bar{U}_i/\partial \tau$ and use the second-order backward finite difference scheme to discretize the physical time step (for convenience, the subscript for cell i is omitted)

$$\frac{\partial U}{\partial \tau} + \frac{(1+\phi)(U^{n+1} - U^n) - \phi(U^n - U^{n-1})}{\Delta t} = R(U^{n+1})/\bar{\Omega}_i$$

where $\phi = 1/2$. For the virtual time, first-order forward finite difference scheme is used

$$\frac{U^{m+1} - U^m}{\Delta \tau} + \frac{(1+\phi)(U^{m+1} - U^n) - \phi(U^n - U^{n-1})}{\Delta t} = R(U^{n+1})/\bar{\Omega}_i$$

where m is the step for inner iteration. And the flux term on the right-hand side is

$$R(U^{n+1}) = -\sum_{j \in N(i)} \widehat{F}_{ij} S_{ij}$$

with $N(i)$ being the face-neighbor cells of cell i. And the flux approximated using the first-order Lax scheme is

$$\hat{F}_{ij} = \frac{1}{2}\left[F_i + F_j - \lambda_{ij}(U_j - U_i)\right], j \in N(i)$$

where λ_{ij} is the spectral radius of the flux Jacobian matrix at the cell face shared by cell i and j,

$$\lambda_{ij} = \lambda_c + \lambda_v S_f^2/\bar{\Omega}_i$$

Thus, the right-hand side term can be linearized as

$$\begin{aligned} R(U^{n+1}) &= R(U^n) + \frac{\partial R}{\partial U}\Delta U \\ &= R(U^n) + \sum_{j \in N(i)} \frac{S_{ij}}{2}\left[\left(J(U_i, \mathbf{n}_{ij}) + \lambda_{ij}\right)\Delta U_i \right. \\ &\quad \left. + \left(J(U_j, \mathbf{n}_{ij}) - \lambda_{ij}\right)\Delta U_j\right]. \end{aligned}$$

Because each control volume is closed, for cell i we have

$$\sum_{j \in N(i)} \frac{S_{ij}}{2} J(U_i, \mathbf{n}_{ij}) \Delta U_i = 0$$

The final equation needed to solve is

$$\left(\bar{\Omega}_i \left(\frac{1}{\Delta \tau} + \frac{(1+\phi)}{\Delta t} \right) + \sum_j \frac{|S_{ji}|}{2} |\lambda_{ji}| \right) \Delta U_i^m + \sum_j \frac{|S_{ji}|}{2} \left(J(U_j, n_{ji}) - |\lambda_{ji}| \right) \Delta U_j^m = R_t$$

with the term on the right-hand side

$$R_t = R(U^n) - \frac{(1+\phi)(U_i^m - U_i^n) - \phi(U_i^n - U_i^{n-1})}{\Delta t} \bar{\Omega}_i$$

To avoid the explicit computation of Jacobian matrix, the following approximation is used

$$J \Delta U \approx \Delta F = F(U + \Delta U) - F(U)$$

This kind of problem can be solved using the Lower-Upper Symmetric Gauss-Seidel (LU-SGS) method described below.

Forward sweep:

$$D_i \Delta U_i^{m,*} + \sum_{j \in L(i)} \frac{|S_{ji}|}{2} \left(J(U_j, n_{ji}) - |\lambda_{ji}| \right) \Delta U_j = R_t$$

Backward sweep:

$$D_i \Delta U_i^m + \sum_{j \in U(i)} \frac{|S_{ji}|}{2} \left(J(U_j, n_{ji}) - |\lambda_{ji}| \right) \Delta U_j = -D_i \Delta U_i^{m,*}$$

where $L(i)$ and $U(i)$ represent the lower and upper face-neighbor elements with respect to cell i, and the diagonal matrix is

$$D_i = \bar{\Omega}_i \left(\frac{1}{\Delta \tau} + \frac{(1+\phi)}{\Delta t} \right) + \sum_j \frac{|S_{ji}|}{2} |\lambda_{ji}|$$

Usually the so-called re-order technique is used to speed up convergence on unstructured grids [2], i.e., rearrange the cell numbering order so as to make the index of face-neighbor cells close the central cell. Since our grids are generated using the commercial software with reasonable numbering order, the re-order step is not used.

2.3 High-Order *K*-Exact Reconstruction

To calculate the left and right states on the faces, we need to reconstruct the distributions using the cell-average values, which is also the key of FVM to achieve high-order accuracy. One of the pioneering work in this area is the FV scheme based on the k-exact reconstruction, which was first proposed by Barth and Fredrichson [3] and later was furthered improved to other applications by Delanaye and Liu [4], Ollivier-Gooch [5, 6] and others. The k-exact reconstruction is of several important properties, namely conservation of the mean, k-exactness and compact support [5, 6]. In this paper, the second-, third-, and fourth-order schemes will be studied which are corresponding to k = 1, 2, and 3. Then the conservation of the mean property requires

$$\frac{1}{\bar{\Omega}_i}\int\limits_{\Omega_i} u_i(x)\,\mathrm{d}\Omega = \bar{u}_i \tag{2.9}$$

where $\bar{u}_i$ is the exact cell average of $u(\mathbf{x})$. Let the center of gravity for Ω_i be $\mathbf{x}_i = (x_i, y_i, z_i)$. The mean preserving property can be automatically satisfied if the reconstruction polynomial is expressed as:

$$u_i(x) = \bar{u}_i + \sum_{l=1}^{DOF(k)} u_i^l \varphi_{l,i}(x),\quad DOF(k) = \frac{(k+1)\cdots(k+d)}{d!} - 1, d = 3 \tag{2.10}$$

where the zero-mean basis φ_k can be expressed as:

$$\begin{aligned}
&\varphi_{1,i}(\mathbf{x}) = \delta x_i,\ \ \varphi_{2,i}(\mathbf{x}) = \delta y_i, \varphi_{3,i}(\mathbf{x}) = \delta z_i,\\
&\varphi_{4,i}(\mathbf{x}) = \delta x_i^2 - \overline{\delta x_i^2}\\
&\varphi_{5,i}(\mathbf{x}) = \delta x_i \delta y_i - \overline{\delta x_i \delta y_i}\\
&\varphi_{6,i}(\mathbf{x}) = \delta x_i \delta z_i - \overline{\delta x_i \delta z_i}\\
&\varphi_{7,i}(\mathbf{x}) = \delta y_i^2 - \overline{\delta y_i^2}\\
&\varphi_{8,i}(\mathbf{x}) = \delta y_i \delta z - \overline{\delta y_i \delta z_i}\\
&\varphi_{9,i}(\mathbf{x}) = \delta z_i^2 - \overline{\delta z_i^2}\\
&\ldots\\
&\delta x_i = x - x_i, \delta y_i = y - y_i, \delta z_i = z - z_i,
\end{aligned} \tag{2.11}$$

with the moments defined as

$$\overline{\delta x_i^{m_1} \delta y_i^{m_2} \delta z_i^{m_3}} = \frac{1}{\bar{\Omega}_i}\int\limits_{\Omega_i} (x - x_i)^{m_1}\,(y - y_i)^{m_2}(z - z_i)^{m_3}\,\mathrm{d}x\mathrm{d}y \tag{2.12}$$

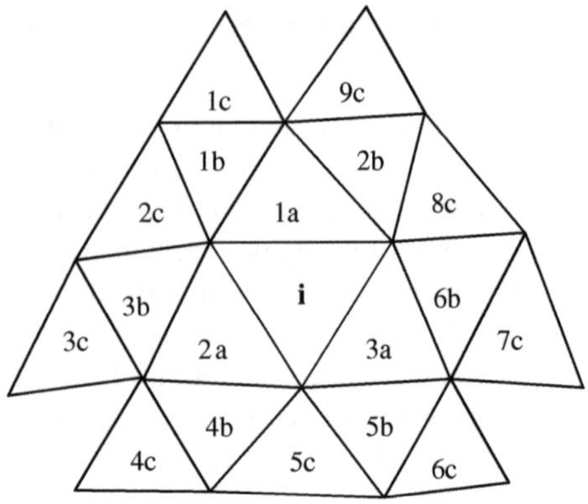

Fig. 2.2 Two-dimensional reconstruction stencils. Second order, $S_i = i, 1a, 2s, 3a$; their order, $S_i = \{i, 1a, 2a, 3a, 1b, 2b, \cdots, 6b\}$; fourth order, $S_i = \{i, 1a, 2a, 3a, 1b, 2b, \cdots, 6b, 1c, 2c, \cdots, 6c\}$. Reprinted from Ref. [7], Copyright 2012, with permission from Elsevier

The property of k-exactness means that $u_i(\mathbf{x})$ reconstructs polynomials of degree k or less exactly, which can be used to determine the coefficients u_i^l. In the neighborhood of Ω_i, there exists a degree k polynomial $P(\mathbf{x})$ that approximates the exact solution $u(\mathbf{x})$ to $O(h^{k+1})$. Because of the k-exactness property of the solution, $u_i(\mathbf{x})$ reconstructs $P(\mathbf{x})$ exactly. Therefore, in the neighborhood of Ω_i represented by a local stencil S_i, we have

$$\frac{1}{\bar{\Omega}_j}\int\limits_{\Omega_j} u_i(\mathbf{x})\,\mathrm{d}\Omega = \frac{1}{\bar{\Omega}_j}\int\limits_{\Omega_j} P(\mathbf{x})\,\mathrm{d}\Omega = \bar{u}_j + O(h^{k+1}), \quad \forall j \in S_i \tag{2.13}$$

where h is the length scale which can be chosen as the maximum radius of the circumcircle of the control volumes, or just be simply defined as $h = \bar{\Omega}^{1/3}$ for non-distorted element. Eq. (2.13) is a system of linear equations of the coefficients u_i^l. To avoid possible singularity, the number of cells in the stencil is required to be larger than the number of unknowns which makes the system over-determined. This system is solved using the method of least squares based on the singular value decomposition.

The left problem is the choosing for the reconstruction stencils S_i. The compact support property implies that the cells in the stencil will not be topologically far from control volume Ω_i. To ensure the systematic and efficient implementation of the k-exact reconstruction, the reconstruction stencil should be as compact as possible. In the present paper, starting from Ω_i, we recursively add the direct face (Von Neumann) neighbors of the cells that are already included in the reconstruction stencil until there is sufficient number of cells in the stencil. Specifically speaking, starting from the central cell as shown in Figs. 2.2 and 2.3, we choose the face-neighbor cells as the first reconstruction stencils, and then take them as the outer border to choose face-neighbor cells as the second reconstruction stencils. The third and fourth reconstruction can be chosen in a similar way.

Fig. 2.3 Three-dimensional reconstruction stencils

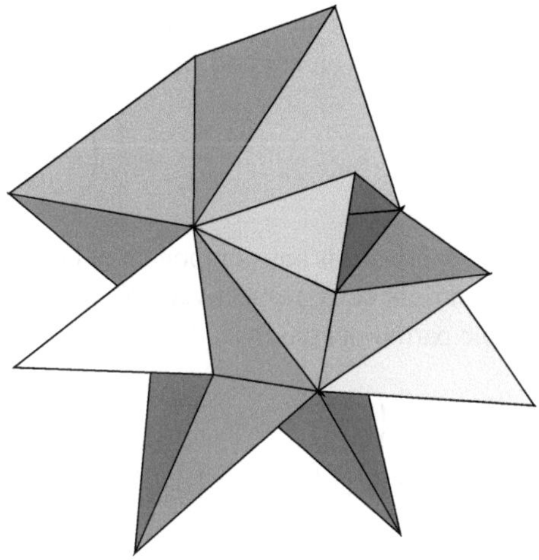

2.4 Evaluation of Viscous Flux in High-Order Accuracy

In the second-order FVM, we assume the linear distribution for the physical quantities, thus the gradient are constant inside every cell. The evaluation of viscous flux is always first-order accurate, which is not adequate to compute the viscous flow dominated by the boundary layer and vortex structures. In the high-order k-exact reconstruction, the distribution of physical data can be also high order accuracy and brings the possibility to compute the viscous flux in high-order accuracy. Now few references talk about the high-order accuracy viscous flux evaluation. Olliver [5] used the central difference like scheme to compute the viscous flux in the convective-diffusive equation, and Titarev [8] used the same way to compute the viscous flux in Navier-Stokes equations by carrying out the k-exact reconstruction for the terms in viscous flux, such as the gradient of velocity and temperature. In the following sections, we will provide a simple yet an efficient way to compute the viscous flux for the Navier-Stokes equations. This method takes use of the original k-exact reconstruction for the conservative variables to compute the viscous flux directly, and thus reduce the times of reconstruction greatly. The boundary conditions are applied on the k-exact reconstructions through the constrained conditions, which are solved using the constrained least-square methods.

2.4.1 The Viscous Flux Computation

To compute the viscous flux of the face, the gradients, $\partial v_i/\partial x_j$, $\partial T/\partial x_j$, are required to be computed first. Since we only know the reconstructed polynomials of conservative variables, these gradients on the face can be computed by the following two steps.

First, the gradient of conservative variables is computed by simple arithmetic average of the gradients on left and right sides, i.e.,

$$\frac{\partial q}{\partial x_j} = \frac{1}{2}\left(\frac{\partial q_L(\mathbf{x})}{\partial x_j} + \frac{\partial q_R(\mathbf{x})}{\partial x_j}\right) \tag{2.14}$$

where q represents any component of the conservative variables $\bar{U}$, L and R represent left and right cells, respectively. Next, the required gradients are computed by some simple partial operations.

$$\begin{aligned}
\frac{\partial v_i}{\partial x_j} &= \frac{1}{\rho^2}\left(\rho\frac{\partial \rho v_i}{\partial x_j} - \rho v_i\frac{\partial \rho}{\partial x_j}\right),\\
\frac{\partial p}{\partial x_j} &= (\gamma-1)\left(\frac{\partial E}{\partial x_j} - \frac{1}{2}\left(u\frac{\partial \rho u}{\partial x_j} + \rho u\frac{\partial u}{\partial x_j} + v\frac{\partial \rho v}{\partial x_j} + \rho v\frac{\partial v}{\partial x_j} + w\frac{\partial \rho w}{\partial x_j} + \rho w\frac{\partial w}{\partial x_j}\right)\right),\\
\frac{\partial T}{\partial x_j} &= \frac{1}{\rho^2 R}\left(\frac{\rho\partial p}{\partial x_j} - \frac{p\partial \rho}{\partial x_j}\right).
\end{aligned} \tag{2.15}$$

As the required gradients for the velocity and temperature are computed, the viscous flux can be obtained by

$$\oint_{\partial\Omega_i} \mathbf{F}_v \cdot \mathbf{n}\,\mathrm{d}S = \sum_{k=1}^{N_F}\int_{\partial\Omega_{i,k}} \mathbf{F}_v \cdot \mathbf{n}\,\mathrm{d}S \approx \sum_{k=1}^{N_F}\left(\sum_{\phi=1}^{N_G}\omega_g \widehat{F}_v(U(\mathbf{x}_{k,\phi}), \nabla U(\mathbf{x}_{k,\phi}))\cdot \mathbf{n}S_k\right)$$

And the values of conservative variables take the mean values on the face.

2.4.2 *On the Implementation of Viscous Boundary Conditions*

In this part we enforce the Dirichlet, Neumann-type boundary conditions by constraining the least-squares reconstruction in control volumes adjacent to the boundary. The constrained conditions are the boundary conditions on the Gauss quadrature points.

$$u(\mathbf{x})|_{\mathbf{x}_g} = u_g,\ g = 1, \cdots, N_g$$

where u_s is the boundary condition of variable u on the quadrature points. Substitute the reconstruction polynomial to the above formula and get

$$\bar{u}_i + \sum_{l=1}^{K} u_i^l \varphi_{l,i}(\mathbf{x}_g) = u_g$$

By collecting the constrained condition equations, we have

$$\begin{aligned}\mathbf{C}\mathbf{u}_i &= \mathbf{d} \\ \mathbf{C} &= \left[\varphi_{l,i}(\mathbf{x}_g)\right], \mathbf{d} = \left[u_g - \bar{u}_i\right] \\ l &= 1, \cdots, K, \quad g = 1, \cdots, N_g\end{aligned} \tag{2.16}$$

And in the k-exact reconstruction, we already have

$$\mathbf{A}\mathbf{u}_i = \mathbf{b} \tag{2.17}$$

The problem becomes the over-determined system (2.17) with constrained conditions (2.16), which can be solved using the constrained least-square method introduced below. By introduce the Lagrange multiplier λ, the constrained least-square methods for solving the above equation are to make the following residual minimize

$$f = \sum_{i=1}^{Ns}\left[\sum_{j=1}^{K} a_{ij}x_j - b_i\right]^2 - \sum_{i=1}^{Ng}\lambda_i\left[\sum_{j=1}^{K} c_{ij}x_j - d_i\right]$$

where a_{ij} and c_{ij} are the components of $\mathbf{A}$ and $\mathbf{C}$, b_i, d_i and x_i are the components of vector $\mathbf{b}$, $\mathbf{d}$ and $\mathbf{u_i}$, Ns is the number of reconstruction stencils. Differentiations for the residual f yields

$$\frac{\partial f}{\partial x_i} = 0, \quad \frac{\partial f}{\partial \lambda_j} = 0, \quad i = 1, \cdots, K, \; j = 1, \cdots, N_g$$

and final equations for computed $\mathbf{x}$, λ becomes

$$\begin{bmatrix} 2\mathbf{A}^T\mathbf{A} & -\mathbf{C}^T \\ \mathbf{C} & 0 \end{bmatrix}\begin{bmatrix} \mathbf{u}_i \\ \lambda \end{bmatrix} = \begin{bmatrix} 2\mathbf{A}^T\mathbf{b} \\ \mathbf{d} \end{bmatrix}$$

To save the computational effort, the inverse of left matrix and matrix A are stored. Actually we only need matrix and vector multiplication operator to compute the coefficients $\mathbf{u}_i$.

2.5 Parallelization for Large-Scale Computation

In the practical engineering applications, large-scale grids are required to make the calculation more accurate, which always need to work on several computers simultaneously. In the computation, we need to divide the computational grids into partitions and one process is responsible for one partition. The most challenge is the data exchange in the interfaces of partitions. Comparing to the second-order scheme, the parallelization for the high-order FVM is more difficult since we need to create a

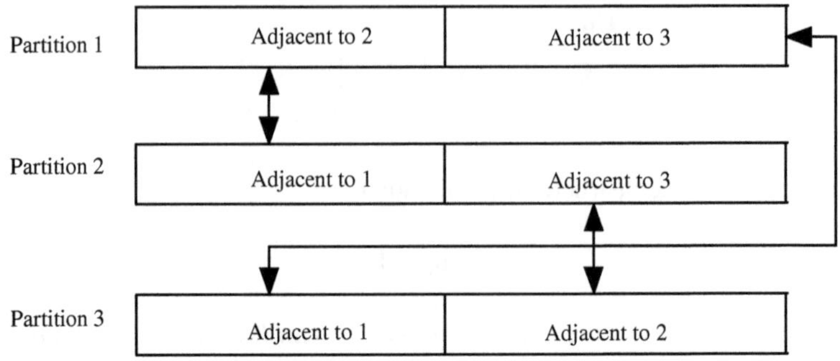

Fig. 2.4 Data transfer model for the parallel program

large set of virtual grids related to the adjacent partitions and transfer much larger data on these regions.

First we use the METIS software [9] to divide the computational grids into several partitions and obtain the partition index for each cell. To exchange data between two adjacent partitions, virtual grids large enough are setup to store the data of neighbor partitions. This is to make the interfacial cells work exactly as the inner cells. According to the algorithms, the data exchange is required on the k-exact reconstruction and flux evaluation. Especially the k-exact reconstruction needs data on large stencils. With the partition information, the interfacial cells can be identified. We create the virtual grids for the reconstruction stencils of the interfacial cells, which are on different partitions from that of the interfacial cells.

With these virtual grids, data exchange must be carried on during the computation. According to the numerical algorithm, the mean values of physical data are exchanged before the k-exact reconstruction and the reconstruction polynomials are exchanged to compute the left and right state used for the numerical flux evaluation. The data exchange is realized using the Message Passing Interface (MPI) standard library. As shown in Fig. 2.4, every partition is adjacent to several other partitions, thus the MPI non-standard message transfer model is used, i.e., every partition firstly sends all data to target partitions, then receive the data from other partitions and set the values on the virtual grids.

2.6 Numerical Tests

This section will solve the two- and three-dimensional Navier-Stokes equations using second- to fourth-order FVMs. The accuracy of high-order FVM will be validated first, then a series of compressible viscous flows are used to verify the correctness of the schemes.

Fig. 2.5 Computational grids (*left*) and initial conditions (*right*) for the two-dimensional convective problem

2.6.1 Accuracy Validation

The following two standard tests are carried out to test the accuracy of the high-order FVM.

2.6.1.1 Two-Dimensional Linear Convective Equation

The two-dimensional linear convective equation can be written as

$$u_t + \mathbf{a} \cdot \nabla u = 0$$

where $\mathbf{a}$ is the constant convective velocity taken as $(1, 2)$ in this test. The convective flux is calculated using the upwind schemes. The initial wave is taken as

$$u_0(x, y) = \sin(2\pi x)\sin(2\pi y)$$

The left picture of Fig. 2.5 shows the computational grids on region $[0, 1] \times [0, 1]$ with grid size $h = 1/20$. The boundaries are periodic. The right picture of Fig. 2.5 shows the computational results of fourth order FVM at $t = 1$ on $h = 1/40$ grids, which return to the initial values. A series of successive refined grids are used to assess the accuracy of the high-order FVM, which are shown in Table 2.2. From this table we can observe that all the second- to fourth-order schemes can achieve the expected orders.

Table 2.2 Accuracy test for the linear convective equation

Scheme	Grid	L1 error	Order	L error	Order
2nd order	1/20	1.14E-02		4.63E-02	
	1/40	2.56E-03	2.16	1.16E-02	2
	1/80	6.00E-04	2.09	2.80E-03	2.05
	1/160	1.39E-04	2.11	8.07E-04	1.8
	1/320	3.55E-05	1.97	2.62E-04	1.62
3rd order	1/20	2.07E-02		3.73E-02	
	1/40	2.69E-03	2.95	4.81E-03	2.95
	1/80	3.37E-04	2.99	6.01E-04	3
	1/160	4.22E-05	3	7.50E-05	3
	1/320	5.27E-06	3	9.37E-06	3
4th order	1/20	1.26E-03		2.67E-03	
	1/40	7.51E-05	4.06	1.56E-04	4.1
	1/80	4.64E-06	4.02	9.33E-06	4.06
	1/160	2.89E-07	4	5.71E-07	4.03
	1/320	1.82E-08	3.99	3.57E-08	4

2.6.1.2 The Constructed Solutions for the Three-Dimensional Euler and Navier-Stokes Equations

These two solutions are used to test the accuracy of the three-dimensional high-order FVM. These are unsteady cases. Give an unsteady vector solutions $U_p(\mathbf{x}, t)$ and substitute them into the right-hand side of the equations,

$$\frac{\partial U}{\partial t} + \frac{\partial \left(F - F_v\right)}{\partial x} + \frac{\partial \left(G - G_v\right)}{\partial y} + \frac{\partial \left(H - H_v\right)}{\partial z} = S.$$

Next we will define two sets of initial flow fields and the corresponding source terms, for which the governing equations are the Euler and Navier-Stokes equations. The computational grids are shown in Fig. 2.6.

For the inviscid Euler equations, the exact solution and the corresponding source terms are defined as

$$U_p(\mathbf{x}, t) = \begin{pmatrix} \rho \\ \mathbf{V} \\ p \end{pmatrix} = \begin{pmatrix} 2 + A_0 \sin(\omega t - \mathbf{kx}) \\ \mathbf{0} \\ 2 + A_0 \sin(\omega t - \mathbf{kx}) \end{pmatrix}$$

and

$$\mathbf{S} = \begin{pmatrix} S_1 \\ S_2 \\ S_3 \\ S_4 \\ S_5 \end{pmatrix} (\mathbf{x}, \mathbf{t}) = \begin{pmatrix} \omega A_0 \cos(\omega t - \mathbf{kx}) \\ -k_x A_0 \cos(\omega t - \mathbf{kx}) \\ -k_y A_0 \cos(\omega t - \mathbf{kx}) \\ -k_z A_0 \cos(\omega t - \mathbf{kx}) \\ \frac{1}{\gamma - 1} \omega A_0 \cos(\omega t - \mathbf{kx}) \end{pmatrix}$$

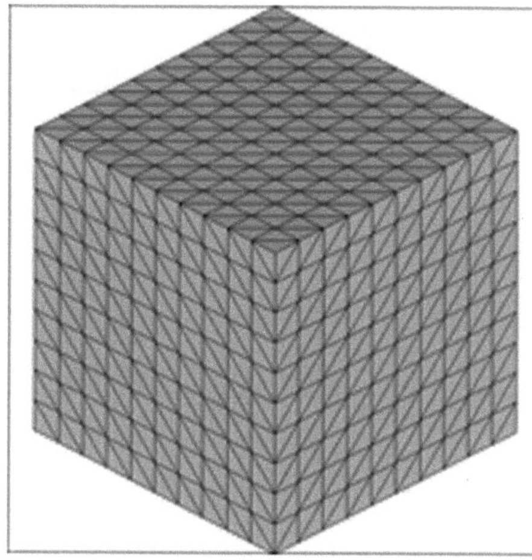

Fig. 2.6 The three-dimensional computational grids for accuracy test

Table 2.3 The accuracy test for the three-dimensional Euler equations

Scheme	Grid	L1 error	Order	L error	Order
3rd order	1/10	3.28E-03		9.84E-03	
	1/20	4.10E-04	3	1.26E-03	2.96
	1/30	1.22E-04	3	3.71E-04	3.02
	1/40	5.13E-05	3	1.56E-04	3
4th order	1/10	2.15E-04		4.07E-04	
	1/20	4.27E-05	3.99	8.34E-05	3.91
	1/30	1.35E-05	4.01	2.65E-05	3.99
	1/40	5.51E-06	4.01	1.08E-05	4.01

where the constant $A_0 = 1$ and $\mathbf{k} = k_x, k_y, k_z, k_x = k_y = k_z = \omega = 2\pi$. The computational time is $t = 0.25$. The 3rd and 4th order FV schemes are used and the error using L_1 and L_∞ are shown in Table 2.3, where we can find the convergence speeds of errors can achieve the expected ones.

For the viscous flows, the exact solution are defined as

$$U_p(\mathbf{x}, t) = \begin{pmatrix} \rho \\ \mathbf{v} \\ p \end{pmatrix} = \begin{pmatrix} 1 \\ \mathbf{0} \\ 2 + A_0 \sin(\omega t - \mathbf{kx}) \end{pmatrix}$$

with temperature

$$T = p/\rho = 2 + A_0 \sin(\omega t - \mathbf{kx})$$

where the constant $A_0 = 1$ and $\mathbf{k} = k_x, k_y, k_z, k_x = k_y = k_z = \omega = 2\pi$. The computational time is $t = 0.25$. The viscous viscosity is taken as $\mu = 1$ and Prantle number as 1. The source terms are

Table 2.4 The accuracy test for the three-dimensional Navier-Stokes equations

Scheme	Grid	L_1 error	Order	L_∞ error	Order
3rd order	1/10	3.68E-03		7.12E-03	
	1/20	7.00E-04	2.4	9.96E-04	2.84
	1/30	2.87E-04	2.2	4.30E-04	2.07
	1/40	1.57E-04	2.11	2.39E-04	2.05
4th order	1/10	4.68E-04		8.05E-04	
	1/20	9.92E-05	3.83	1.79E-04	3.71
	1/30	3.03E-05	5.39	5.57E-05	5.23
	1/40	1.18E-05	4.2	2.25E-05	4.07

$$S = \begin{bmatrix} 0 \\ -k_x A_0 \cos(\omega t - \mathbf{kx}) \\ -k_y A_0 \cos(\omega t - \mathbf{kx}) \\ -k_z A_0 \cos(\omega t - \mathbf{kx}) \\ \frac{1}{\gamma - 1}\omega A_0 \cos(\omega t - \mathbf{kx}) + k_{\text{conduct}} A_0 \sin(\omega t - \mathbf{kx}) \left(k_x^2 + k_y^2 + k_z^2\right) \end{bmatrix}$$

The 3rd and 4th order FV schemes are used and the error using L_1 and L_∞ are shown in Table 2.4, where we can find the convergence speeds of errors can achieve the expected ones.

2.6.2 The Flat Plate Boundary Layer

We consider the laminar boundary over an adiabatic flat plate at a free-stream Mach number of 0.3 and a Reynolds number of 15,000 based on the free-stream velocity and the length of the flat plate. The computation domain is $[-0.03, 0.1] \times [0., 0.03]$ with 24,700 element, as shown in Fig. 2.7. The flat plate lies from $x = 0$ to 0.1.

The second- to fourth-order FV schemes are used in the computation. The u&v nondimensional velocities in the normal directions are shown in Fig. 2.8, where we can observe that the second- to fourth-order results can agree with the exact Blasius solutions quite well. This test case validates the correctness of using the mean values of conservative variables to compute the viscous flux and the constrained least-square method to treat the viscous boundary conditions.

2.6.3 The Subsonic Flow Around NACA0012 Airfoil

This test case deals with a subsonic flow past a NACA0012 airfoil at a Mach number of 0.5, and an angle of attack 0°, and a Reynolds number of 5,000 based on the

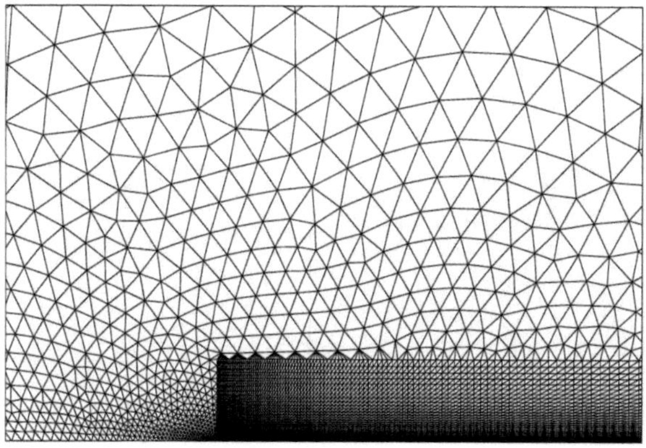

Fig. 2.7 Regions and grids for the boundary layer plate

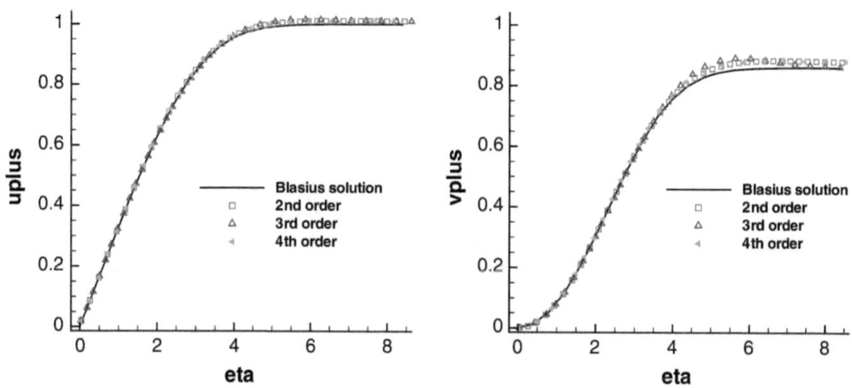

Fig. 2.8 The u and v velocity distribution in the normal directions

free-stream velocity and the chord length of the airfoil. An adiabatic wall is assumed in this case. The computational grids are shown in Fig. 2.9.

We use a well-tested second-order scheme to compute the pressure and skin friction coefficient on a fine grids and take the results as the reference values. The Mach number contour is shown in the right side of Fig. 2.9. The 3rd and 4th order schemes are used to compute the pressure and skin friction distributions. The results are shown in Fig. 2.10, where the coefficients can match the reference values quite well.

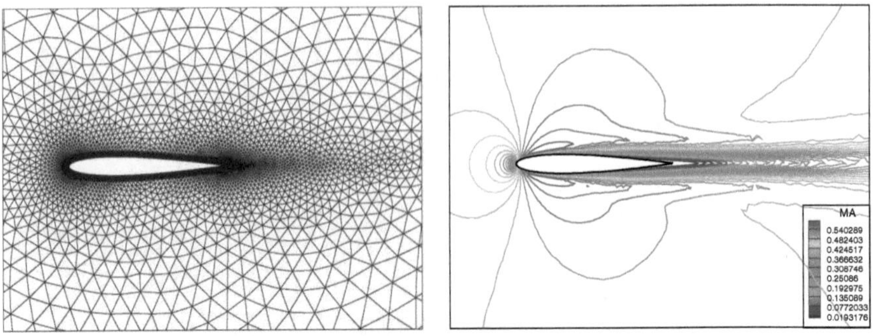

Fig. 2.9 Grid and mach number contour for NACA0012

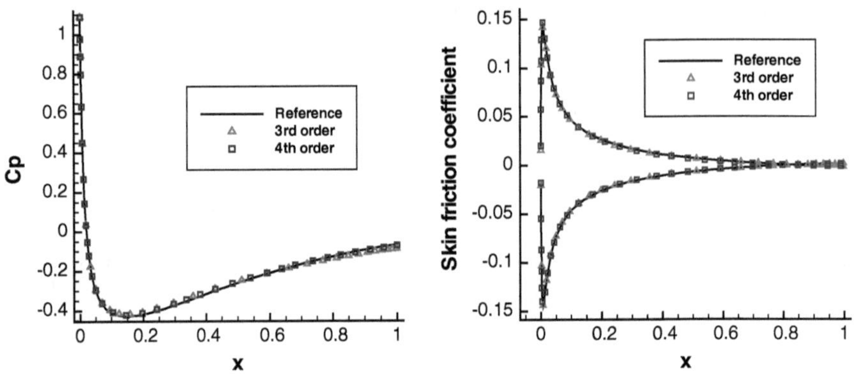

Fig. 2.10 Grid and Mach number contour for NACA0012

2.6.4 Sound Field Generated by the Flow Around a Circular Cylinder

This test case concerns about the sound field generated by the unsteady Kármon street vortex. We have the cylinder diameter $d = 1$, the Reynolds number based on the cylinder diameter and free stream parameters $\text{Re} = \rho_\infty u_\infty d/\mu_\infty = 150$ and the inlet Mach number $M_\infty = 0.2$. Since the pressure fluctuation computation requires high accuracy numerical schemes, the 4th order FV scheme is used in the computation where the curved boundary is approximated using 3rd order polynomials for each segment. The grids are shown in the left of Fig. 2.11 with 83,936 triangular grids. The sound field caused by the Kármon street vortex is shown in the right of Fig. 2.11, which agrees with the results of Müller [10] and Dumbser [11] quite well. And the Mach contour in the vortex is shown in Fig. 2.12. The historical lift and drag coefficients are shown in Fig. 2.13 and the corresponding Strouhal number can be computed $St = fd/\mu_\infty = 0.182$, which agrees with the results of Müller [10] and Dumbser [11] $St = 0.182$ quite well.

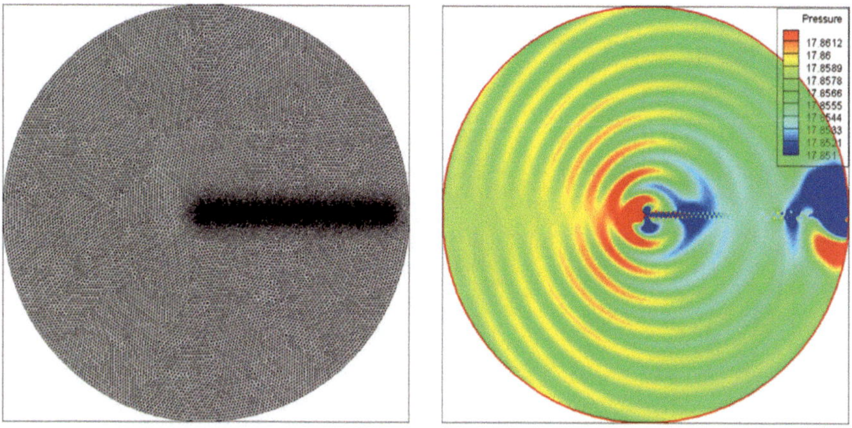

Fig. 2.11 Grid and Mach number contour for NACA0012

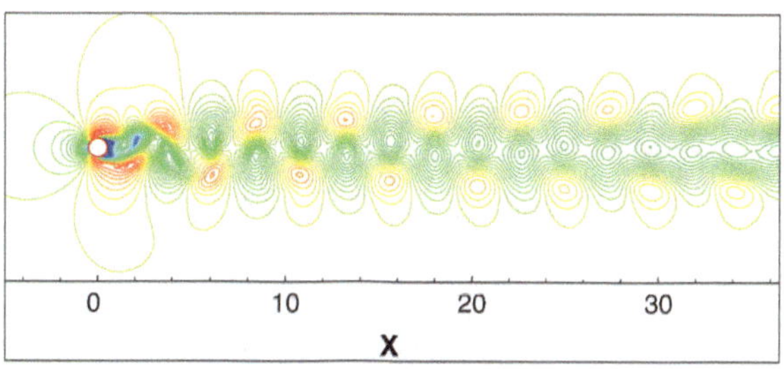

Fig. 2.12 The Mach contour line in the flow around cylinder

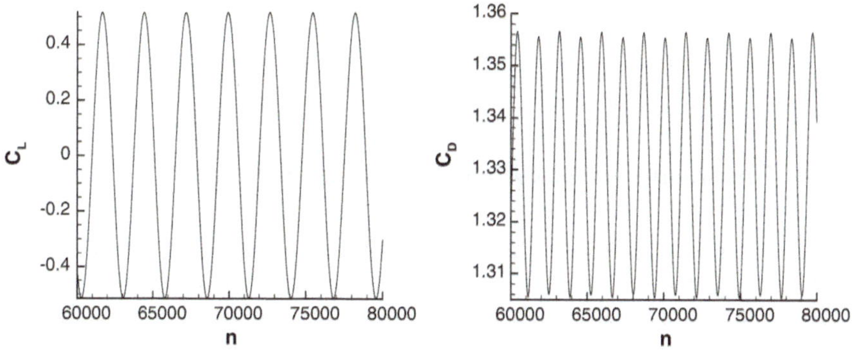

Fig. 2.13 Grid and Mach number contour for NACA0012

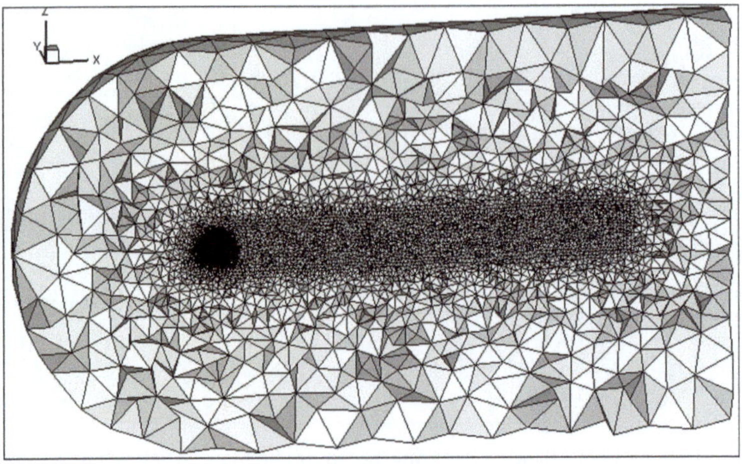

Fig. 2.14 The grids for computing viscous flows around sphere

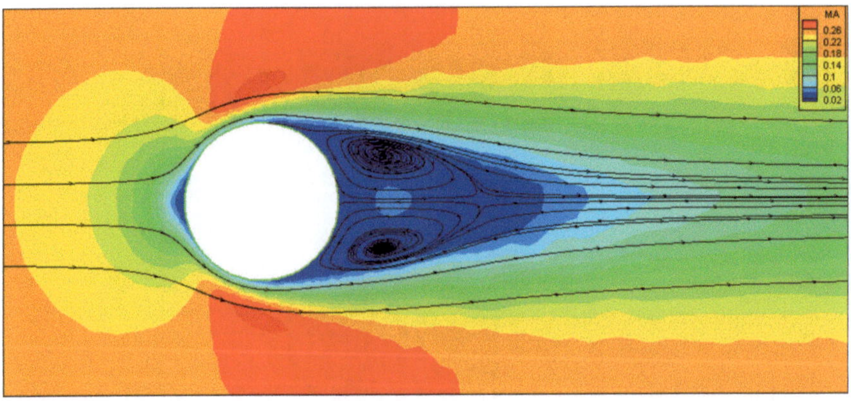

Fig. 2.15 The Mach number contour with stream lines at $z = 0$ plane computed using 4th order FVM

2.6.5 The Steady Viscous Flow Around the Sphere

To validate the correctness of high-order FVM in solving the three-dimensional compressible flows, this section is to compute the steady viscous flow around the sphere. The experimental and computational conditions are taken from reference [12, 13] for comparison. The Reynolds number based on the spherical diameter is 118 with the inlet Mach number 0.2535. The spherical diameter is 1 and the computational region is combined by a half sphere of diameter 10 and a cylinder of height 25. The computational grids are shown in Fig. 2.14 with 458,915 tetrahedrons and 77,844 grid points. The mesh size near the sphere is taken as $h = 0.04$. The 3rd and 4th

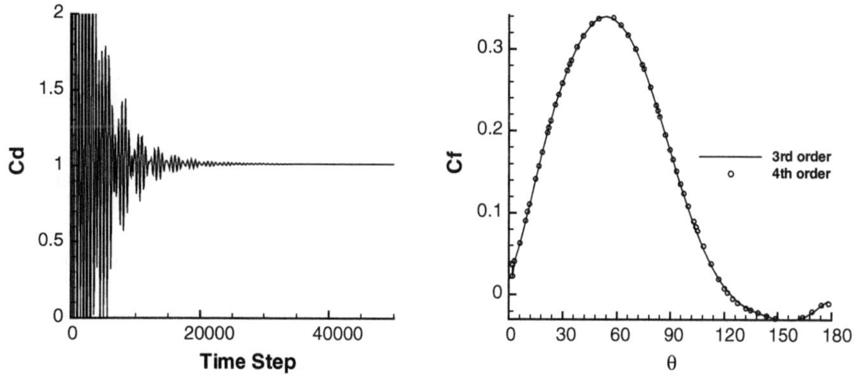

Fig. 2.16 The historical drag coefficients (*left*) and the skin friction coefficients (*right*, θ is angle)

order FV schemes are used in the computations. The Mach number and stream lines can be clearly observed in the wake of the sphere shown in Fig. 2.15. And Fig. 2.16 shows the converged drag coefficient $C_d = 1.01$, which agrees with the results 1.0 in Refs. [12, 13] quite well. Also the order-convergence skin friction coefficients using the 3rd and 4th order FV schemes agree with the Ref. [12] well.

2.7 Conclusions

This chapter completes the framework of high-order finite volume schemes in solving compressible Navier-Stokes equations. Based on the k-exact reconstruction, the high-order quadrature on the faces is carried out using the Gaussian quadrature rules and the FVM is generalized to high-order accuracy. To compute the diffusive flux in high-order accuracy, the key element is the evaluation of the gradients on the face, which is chosen to be the mean value of gradient of reconstruction polynomials on left and right cells. This procedure is very efficient and does not bring in additional reconstruction steps. Also we consider the viscous boundary condition by incorporating it into the k-exact reconstruction step through the constrained least-square method. Two and three-dimensional numerical results show that the high-order scheme can predict the viscous flow accurately.

References

1. Jameson A, Yoon S (1987) Lower-upper Implicit schemes with multiple grids for the Euler equations. AIAA J 25:929–935
2. Sharov D, Nakahashi K (1997) Reordering of 3D hybrid unstructured grids for vectorized LU-SGS Navier-Stokes computations. AIAA Paper 97–2102

3. Barth TJ, Frederichson PO (1990) Higher order solution of the Euler equations on unstructured grids using quadratic reconstruction. AIAA 90–0013
4. Delanaye M, Liu Y (1999) Quadratic reconstruction finite volume schemes on 3D arbitrary unstructured polyhedral grids. AIAA 99–3259
5. Ollivier-Gooch C, Altena MV (2002) A high-order-accurate unstructured mesh finite-volume scheme for the advection-diffusion equation. J Comput Phys 181:729–752
6. Ollivier-Gooch CF (1997) Quasi-ENO schemes for unstructured meshes based on unlimited data-dependent least-square reconstruction. J Comput Phys 133:6–17
7. Li WA, Ren YX (2012) The multi-dimensional limiters for solving hyperbolic conservation laws on unstructured grids II: extension to high order finite volume schemes. J Comput Phys 231(11):4053–4077
8. Titarev VA, Drikakis D (2011) Uniformly high-order schemes on arbitrary unstructured meshes. Comput Fluids 46(1):467–471
9. Karypis G, Kumar V (1998) Multilevel k-way partitioning scheme for irregular graphs. J Parallel Distrib Comput 48:96129
10. Müller B (2008) High order numerical simulation of aeolian tones. Comput Fluids 37:450–462
11. Dumbser M (2011) Arbitrary high order PNPM schemes on unstructured meshes for the compressible Navier-Stokes equations. Comput Fluids 39:60–76
12. Haga T, Gao H, Wang ZJ (2010) A high-order unifying discontinuous formulation for 3D mixed grids. AIAA 2010–540
13. Taneda S (1956) Experimental investigations of the wake behind a sphere at low Reynolds numbers. J Phys Soc Jpn 11:1104–1108

Chapter 3
Accuracy Preserving Limiters for High-Order Finite Volume Methods

One of the bottlenecks for high-order numerical methods on unstructured grids is the design of effective and robust limiters for the simulation of flow with discontinuities. The challenge is to maintain the high- order accuracy in smooth regions while capturing the discontinuities in high resolutions without oscillations. This chapter will provide several high order and high resolution nonoscillatory limiters for the high-order FVM. Before introducing the unstructured FVM limiter, we review the historical research for the nonoscillatory schemes in the development of computational fluid methods.

To study the properties of equations, we use the basis model equation, i.e., one-dimensional linear scalar equation

$$\frac{\partial u}{\partial t} + a\frac{\partial u}{\partial x} = 0$$

The linear discretization scheme for the model equation can be expressed as

$$u_i^{n+1} = H\left(u_{i-r+1}^n, \ldots, u_{i+s}^n\right) = \sum_{p=-r+1}^{s} c_{i,p} u_{i+p}^n$$

With this scheme, some properties about the nonoscillatory scheme are introduced below.

Monotone. Suppose the scheme is defined as

$$u_i^{n+1} = H\left(u_{i-r+1}^n, \ldots, u_{i+s}^n\right) \tag{3.1}$$

where r and s are positive integers. If $\forall j,\ \ \partial H/\partial u_j^n \geq 0$, we call (3.1) the monotone scheme. The monotone scheme is a nonoscillatory scheme. According to the Godunov theorem [1], any linear scheme higher than first order cannot give monotonic solutions. The high-order schemes (≥2nd order) will generate the non-physical oscillations in the vicinities of discontinuities, which is called the Gibbs

W. Li, *Efficient Implementation of High-Order Accurate Numerical Methods on Unstructured Grids*, Springer Theses, DOI: 10.1007/978-3-662-43432-1_3,

phenomenon. The current high-order schemes always introduce the nonlinear treatments to make the schemes satisfy monotone property.

Total Variation Diminishing (TVD). The scheme

$$u_i^{n+1} = H\left(u_{i-r+1}^n, \ldots, u_{i+s}^n\right) \tag{3.2}$$

can be called the TVD scheme, if satisfying

$$TV(u^{n+1}) \leq TV(u^n),\ \forall n \tag{3.3}$$

where TV is called the total variation to measure the smoothness of function. It is defined on the continuous differential function as

$$TV(u) = \int_{-\infty}^{+\infty} \left|u'(x)\right| dx$$

on scattered points $u^n = \left\{u_i^n\right\}$ as

$$TV(u^n) = \sum_{i=-\infty}^{+\infty} \left|u_{i+1}^n - u_i^n\right|.$$

It is easy to know that the scheme satisfying TVD condition is also monotonicity preserving.

Essentially Non-Oscillatory (ENO) [2]. It means in the high-order reconstruction of smooth function $V(\mathbf{x})$, the piecewise smooth polynomial $P(\mathbf{x})$ is obtained which achieve high-order accuracy in smooth regions and avoid the Gibbs phenomenon near the discontinuities. In fact, ENO is high-order limiting strategy, while the TVD scheme is at most second order accurate [3]. Thus, the ENO scheme does not satisfy the TVD condition. But it meets the total variation bounded (TVB) condition, i.e., there exists a function $z(\mathbf{x})$, if it satisfies the following condition on any elements

$$z(x) = P_i(x) + O(\Delta x^{k+1}), x \in I_i$$

then $TV(z) \leq TV(V)$.

Maximum Principle (MP). The MP can be defined as that if $M = \max_x u_0(x)$ and $m = \min_x u_0(x)$ for the initial value, then $u(x,t) \in [m, M]$ for any x and $t \geq 0$. We can say TVD scheme is also MP satisfying.

Two main approaches are used to capture the discontinuities and avoid the Gibbs phenomenon. The first one is to add artificial viscosities near the discontinuities. A so-called shock detector is used to mark the regions with discontinuities, then the dissipation term with artificial viscosities is added to the equation on these marked regions to suppress the oscillations. This method is easy to implement, but the drawback is that we need parameters to identify the trouble cells and add the dissipation. These parameters are problem-dependent and are crucial to the accuracy of the numerical schemes. Thus, no universal parameters can be provided. Meanwhile, the introduction of positive dissipation terms will decrease the time step and increase the computational cost. And it is hard to prove that the schemes with artificial viscosities satisfy the above nonoscillatory criteria described above.

The other method is the reconstruction and limiting method. The design of limiter always uses the information of cells far away and limits the target polynomial near the discontinuities. The good limiters should be insensitive to the parameters and keep the high-order accuracy in smooth regions while suppressing the oscillations near the discontinuities. The TVD limiter function in the second order TVD-MUSCL scheme is derived using the TVD condition. However, the design of high-order limiters can hardly use the principles described above (there are exceptions, e.g., Ref. [4] uses the MP in choosing parameters). The WENO limiter is proved to satisfy the TVB condition after constructed successfully. Thus, the construction for high-order limiters is quite difficult which needs a lot of numerical experiments. And the analysis using the nonoscillatory principles on unstructured grids is extremely difficult. Usually, we judge the efficiency of the limiting strategy through abundance of numerical cases. This chapter is to use the limiters to make the high-order FVM nonoscillatory.

3.1 The Limiters on Unstructured Grids and the Methodology of Constructing Limiters in This Chapter

Using the principles above, many excellent nonoscillatory schemes are designed on structured grids, e.g., the second order TVD-MUSCL scheme, the high-order ENO/WENO scheme. And these schemes are already applied in the simulation of flows with strong shock waves and complicated structures. Nevertheless, these structured schemes cannot be generalized to unstructured grids straightly. There are some exceptions, e.g., the TVD-MUSCL can be generalized to unstructured grids but it cannot suppress the numerical oscillations efficiently, the ENO/WENO method can be generalized to unstructured grids but the construction of candidate polynomials on biased stencils is quite complex. Some of the current unstructured limiters are designed based on the MP. Among these limiters are the widely used Barth and Jespersen limiter [5], Venkatakrishnan limiter [6], and the MLP limiters [7, 8]. Other types of limiters such as the TVB limiter [9], Hermite WENO limiter [10], and moment limiter [11] are also adopted in the DGM and SV/SD methods.

Summarizing on the successful high-order limiters, we find the three components in the limiters, namely the construction of candidate polynomials, the limiting function and the limiting process. Two classical high-order limiters are analyzed in terms of these three components in the below. The WENO limiter uses the multiple reconstructions to provide the candidate polynomials, the smoothness indicator to compute the WENO weight as the limiting function and the weighted summation for the limiting process. The moment limiter uses the derivatives of lower order coefficients to approximate the high-order derivatives which provide the candidate coefficients, the TVD function as the limiting function and the successive limiting process from high-order coefficients to low order ones as the limiting process. Different limiters not only affect the final computation greatly, but also affect the implementation and computational cost. The main work of this chapter is to make the three components in the limiters simpler and more effectively. In the following sections, we will propose the secondary reconstruction to replace the multiple reconstructions to provide the candidate polynomials, construct a new WBAP function for the limiting function, and adopt the improved successive limiting process to make the limiting stencils compact.

3.2 The Secondary Reconstruction

According to the limiting procedure of Eq. (3.17), the arguments of L must approximate the scaled derivatives to the same order. This means on a given cell, we need several candidate reconstructions that approximate the solution to the same order of accuracy. In the WENO schemes, several candidate stencils must be identified and the reconstruction is performed on each of these stencils. This approach is computationally very expensive. In this chapter, an alternative approximation, namely the SR is adopted. The implementation of SR has been introduced in [12]. In this chapter, the SR is introduced in a slightly different perspective to show its construction more clearly.

The basic idea of SR is to modify the PRs on neighboring cells of Ω_i as the reconstructions on Ω_i . One may wonder if this is possible. In the following, we will show that this is indeed possible and furthermore, the modification is accuracy preserving.

Consider cell Ω_i and one of its neighboring cells Ω_j. We argued that as long as $\Omega_i \in S_j$, the PR on Ω_j is also a k-exact reconstruction on Ω_i. This is a direct result of the k-exactness property of the PR. Assuming there exists a degree k polynomial $P(\mathbf{x})$ that approximates the exact solution to $O(h^{k+1})$ on S_j, the PR on Ω_j recovers $P(\mathbf{x})$ exactly not only on Ω_j, but also on every cell of S_j. Therefore, we have

$$u_j(\mathbf{x}) = u_i(\mathbf{x}) + O(h^{k+1}), \mathbf{x} \in \Omega_i. \tag{3.4}$$

Equation (3.4) suggests that the simplest accuracy preserving SR on Ω_i in terms of $u_j(\mathbf{x})$ is

$$u_{j\to i}(\mathbf{x}) = u_j(\mathbf{x}), \mathbf{x} \in \Omega_i. \tag{3.5}$$

However, we emphasize that $u_{j\to i}(\mathbf{x})$ defined by Eq. (3.5) is not identical to the PR on Ω_i using a biased stencil $S_i^* = S_j$ because of the differences in basis functions and also the weight functions. As a result, generally speaking, Eq. (3.5) does not conserve the mean of Ω_i although it is accuracy preserving, i.e.,

$$\frac{1}{\bar{\Omega}_i}\int\limits_{\Omega_i} u_{j\to i}(\mathbf{x})d\Omega \neq \bar{u}_i.$$

To show this more clearly, for $\mathbf{x} \in \Omega_i$, we rewrite Eq. (3.5)

$$u_{j\to i}(\mathbf{x}) = \bar{u}_j + \sum_{l=1}^{DOF} u_j^l \varphi_{l,j}(\mathbf{x})$$

into the following form

$$u_{j\to i}(\mathbf{x}) = \bar{u}_j + \mathbf{u}_j^T \phi_j \tag{3.6}$$

where

$$\mathbf{u}_j = \left[u_j^1, u_j^2, u_j^3, \ldots\right]^T, \phi_j = \left[\varphi_{1,j}, \varphi_{2,j}, \varphi_{3,j}, \ \ldots\right]^T.$$

ϕ_j in Eq. (3.6) can be expressed in terms of ϕ_i defined by

$$\phi_i = \left[\varphi_{1,i}, \varphi_{2,i}, \varphi_{3,i}, \ \ldots\right]^T$$

which is the vector of the basis functions on Ω_i. According to Eq. (2.11), it is easy to derive that

$$\phi_j = \mathbf{A}\phi_i + \mathbf{b} \tag{3.7}$$

where

$$\mathbf{A} = \begin{bmatrix} 1 & 0 & 0 & \cdots \\ 0 & 1 & 0 & \cdots \\ 2\Delta x & 0 & 1 & \cdots \\ \cdots & \cdots & \cdots & \cdots \end{bmatrix}, \mathbf{b} = \begin{bmatrix} \Delta x \\ \Delta y \\ \Delta x^2 + \overline{\delta x_i{}^2} - \overline{\delta x_j{}^2} \\ \cdots \end{bmatrix} \tag{3.8}$$
$$\Delta x = x_i - x_j, \Delta y = y_i - y_j.$$

Using Eqs. (3.7) and (3.6) can be written as

$$u_{j\to i}(\mathbf{x}) = \bar{u}_j + \mathbf{u}_j^T(\mathbf{A}\phi_i + \mathbf{b}) = (\bar{u}_j + \mathbf{u}_j^T\mathbf{b}) + \mathbf{u}_j^T\mathbf{A}\phi_i. \tag{3.9}$$

A straightforward computation shows

$$\frac{1}{\bar{\Omega}_i}\int\limits_{\Omega_i} u_{j\to i}(\mathbf{x})d\Omega = \bar{u}_j + \mathbf{u}_j^T\mathbf{b}$$

Therefore, $u_{j\to i}(\mathbf{x})$ is not conservative on Ω_i as long as

$$\bar{u}_j + \mathbf{u}_j^T \mathbf{b} \neq \bar{u}_i.$$

On the other hand, the cell average of Eq. (3.4) leads to

$$\frac{1}{\bar{\Omega}_i}\int_{\Omega_i} u_{j\to i}(\mathbf{x})d\Omega = \frac{1}{\bar{\Omega}_i}\int_{\Omega_i} u_i(\mathbf{x})d\Omega + O(h^{k+1})$$

which is equivalent to

$$\bar{u}_j + \mathbf{u}_j^T \mathbf{b} = \bar{u}_i + O(h^{k+1}). \tag{3.10}$$

Equation (3.10) indicates that although $u_{j\to i}(\mathbf{x})$ is not conservative on Ω_i, the conservation error is very small. Equation (3.10) also suggests a simple modification to Eq. (3.9) to restore the conservation, which is to replace $\bar{u}_j + \mathbf{u}_j^T \mathbf{b}$ in Eq. (3.9) with $\bar{u}_i$. Therefore, the practical SR on Ω_i in terms of $u_j(\mathbf{x})$ can be written as

$$u_{j\to i}(\mathbf{x}) = \bar{u}_i + (\mathbf{u}_j^T \mathbf{A})\boldsymbol{\phi}_i. \tag{3.11}$$

Equation (3.11) can be also written in a component-wise form which is more straightforward in implementation

$$u_{j\to i}(\mathbf{x}) = \bar{u}_i + \sum_{l=1}^{DOF} u_{j\to i}^l \varphi_{l,i}(\mathbf{x}) \tag{3.12}$$

where $u_{j\to i}^l$ is the l-th component of $\mathbf{u}_j^T \mathbf{A}$, e.g., for a 4th order reconstruction,

$$\begin{aligned}
u_{j\to i}^1 &= u_j^1 + 2u_j^3\Delta x + u_j^4\Delta y + 3u_j^6\Delta x^2 + 2u_j^7\Delta x\Delta y + u_j^8\Delta y^2\\
u_{j\to i}^2 &= u_j^2 + u_j^4\Delta x + 2u_j^5\Delta y + u_j^7\Delta x^2 + 2u_j^8\Delta x\Delta y + 3u_j^9\Delta y^2\\
u_{j\to i}^3 &= u_j^3 + 3u_j^6\Delta x + u_j^7\Delta y\\
u_{j\to i}^4 &= u_j^4 + 2u_j^7\Delta x + 2u_j^8\Delta y\\
u_{j\to i}^5 &= u_j^5 + u_j^8\Delta x + 3u_j^9\Delta y\\
u_{j\to i}^l &= u_j^l, l = 6, 7, 8, 9.
\end{aligned} \tag{3.13}$$

the SRs for fourth order FVM on the three-dimensional grids can be written as

$$\begin{aligned}
u_{j\to i}^1 &= u_j^1 + 2u_j^4\Delta x + u_j^5\Delta y + u_j^6\Delta z + 3u_j^{10}(\Delta x)^2 + 2u_j^{11}\Delta x\Delta y + u_j^{12}(\Delta y)^2 +\\
&\quad u_j^{13}\Delta y\Delta z + 2u_j^{14}\Delta x\Delta z + u_j^{15}(\Delta z)^2\\
u_{j\to i}^2 &= u_j^2 + 2u_j^7\Delta y + u_j^5\Delta x + u_j^8\Delta z + u_j^{11}(\Delta x)^2 + 2u_j^{12}\Delta x\Delta y + u_j^{13}\Delta x\Delta z +\\
&\quad 3u_j^{16}(\Delta y)^2 + 2u_j^{17}\Delta y\Delta z + u_j^{18}(\Delta z)^2\\
u_{j\to i}^3 &= u_j^3 + 2u_j^9\Delta z + u_j^6\Delta x + u_j^8\Delta y + u_j^{13}\Delta x\delta y + 2u_j^{14}(\Delta x)^2 + 2u_j^{15}\Delta x\Delta z +
\end{aligned}$$

$$
\begin{aligned}
&u_j^{17}(\Delta y)^2 + 2u_j^{18}\Delta y \Delta z + 3u_j^{19}(\Delta z)^2 \\
u_{j\to i}^4 &= u_j^4 + 3u_j^{10}\Delta x + u_j^{11}\Delta y + u_j^{14}\Delta z \\
u_{j\to i}^5 &= u_j^5 + 2u_j^{11}\Delta x + 2u_j^{12}\Delta y + u_j^{13}\Delta z \\
u_{j\to i}^6 &= u_j^6 + 2u_j^{14}\Delta x + u_j^{13}\Delta y + 2u_j^{15}\Delta z \\
u_{j\to i}^7 &= u_j^7 + u_j^{12}\Delta x + 3u_j^{16}\Delta y + u_j^{17}\Delta z \\
u_{j\to i}^8 &= u_j^8 + u_j^3\Delta x + 2u_j^7\Delta y + 2u_j^8\Delta z \\
u_{j\to i}^9 &= u_j^9 + u_j^{15}\Delta x + u_j^{18}\Delta y + 3u_j^{19}\Delta z \\
u_{j\to i}^{ip} &= u_j^{ip},\ ip = 10, \ldots, 19 \\
\Delta x &= x_i - x_j,\ \Delta y = y_i - y_j,\ \Delta z = z_i - z_j
\end{aligned}
\tag{3.14}
$$

Equations (3.12) and (3.13) show that the SR can be constructed analytically. Therefore, the SR is very efficient, in fact, much more efficient than performing a new reconstruction on a manually selected stencil.

3.3 The Quadrature-Free K-Exact WENO Limiters

The WENO scheme gains great success on structured grids. For the unstructured meshes, there are two types of WENO schemes designed. In the first type of WENO schemes (referred to as WENO-1 hereafter) [13–15], the order accuracy of the WENO reconstruction is the same as that of the reconstruction on each candidate stencil. In the second type of WENO schemes (referred to as WENO-2 hereafter) [16], the weighted average of lower order candidate polynomials obtained by the reconstructions on "small" stencils is designed to achieve higher order of accuracy by using a high-order reconstruction on a "big" stencil to determine the linear weights. However, there is an obvious defect for all these two types of WENO schemes, i.e., the complicated multiple reconstructions on the biased stencils. Besides the ambiguity and complexity in choosing multiple stencils (especially the biased stencils), this procedure requires multiple reconstructions, e.g., seven k-exact reconstructions for 2D case in [14], and is therefore complicated and time-consuming.

The WENO schemes always contain the following two parts. (1) One is the construction of the candidate polynomials. The multiple reconstructions are used in the traditional WENO schemes, which requires to construct the polynomials on a series of candidate stencils. (2) The other is the WENO weight process. The WENO weight usually contains the linear and nonlinear parts. The former is related to the accuracy by specifying the importance of each stencil, while the latter is used to suppress the numerical oscillations.

In the following section, we will improve these two parts to make an easily implemented k-exact WENO scheme. The main features for this scheme are that it uses the efficient secondary reconstruction to construct the candidate polynomials and the quadrature-free way to compute the WENO smooth indicator.

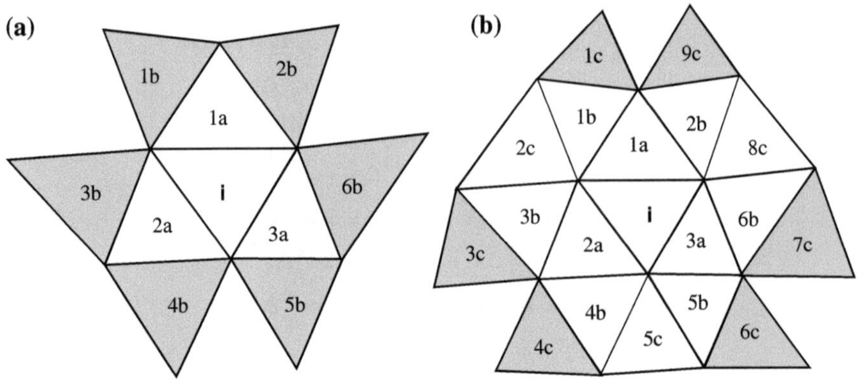

Fig. 3.1 The secondary reconstruction stencils for the third (*left*) and fourth (*right*) order schemes in two-dimensional grids. **a** Triangle **b** Quadrilateral

3.3.1 The Stencils for the Secondary Reconstruction

The stencils for the k-exact WENO schemes are provided through the secondary reconstruction. The choosing method is relatively much easier than the multiple reconstructions which are difficult in choosing the biased stencils [13, 15]. The easiest way is to choose all the stencils in the k-exact reconstruction for the secondary reconstruction. Remember that the candidate polynomials are serving the time-consuming WENO limiting process in the next section, so the number of chosen stencils should be as small as possible and at least one of the stencils is smooth in the presence of discontinuity.

With this objective, the following three principles are made in choosing the stencils:

(1) Cell j in Eq. (3.13) belongs to S_i, i.e., $j \in S_i$. And $i \in S_j$ is also required. The condition guarantees the accuracy of $u_j(\mathbf{x})$ using the SR on cell i.

(2) The stencils should be symmetrical and the farthest away from the central cell topologically.

(3) The stencils should contain every direction with respect to the central cell.

The principles, (2) and (3), are used to assure that at least one oscillation-free candidate polynomial is provided if there's discontinuity on the central cell. For the third and fourth order scheme, the topological distance away from the central cell can be expressed using the number of shared vertex with the central cell. For example, the stencil for the secondary reconstruction on the two-dimensional grids can be chosen as Fig. 3.1, which indicates to choose the cells sharing one and zero vertex with central cells for the third and fourth order schemes, respectively. For other types of orders and grids, the number of shared vertex can be referred to Table 3.1.

Table 3.1 Stencils used for the secondary reconstruction. Number of sharing vertex with the central cell

Type of central cell	Triangle	Tetrahedron
3rd order	1	≤ 2
4th order	0	≤ 1

3.3.2 Quadrature-Free Nonoscillation WENO Reconstruction in Characteristic Space

Using the k-exact and secondary reconstructions, we obtain a series of candidate reconstruction polynomials for cell Ω_i, namely $u_i(\mathbf{x})$ and $u_{j\to i}(\mathbf{x})$, $j \in T_i$. To ease the presentation, these reconstructions are denoted as $u_{i,k}(\mathbf{x})$, $k = 0, \ldots, N_s$, where $u_{i,0}(\mathbf{x})$ represents the polynomial $u_i(\mathbf{x})$, $u_{i,k}(\mathbf{x})$, $k > 0$ represents the polynomials $u_{j\to i}(\mathbf{x})$, N_s is the number of cells in T_i. The final reconstruction polynomial is obtained by the weighted average of these polynomials using the WENO procedure similar to that presented in [14, 15]. Specifically, $u_{i,k}(\mathbf{x})$, $k = 0, \ldots, N_s$ are first transformed to the characteristic space and the corresponding characteristic polynomials are denoted with $v_k(\mathbf{x})$, which can be expressed by

$$v_k(\mathbf{x}) = \bar{v}_i + \sum_{l=1}^{DOF(k)} v_k^l \varphi_{l,i}(\mathbf{x}).$$

Then the smoothness indicators are computed using

$$\begin{aligned}\beta_k &= \sum_{m_1+m_2+m_3=1}^{DOF(k)} h^{2(m_1+m_2+m_3)-3} \int_{\Omega_i} \left(\frac{\partial^{m_1+m_2+m_3} v_k(\mathbf{x})}{\partial x^{m_1} \partial y^{m_2} \partial z^{m_3}}\right)^2 d\Omega \\ &= \sum_{m_1+m_2+m_3=0}^{DOF(2(k-1))} A_{m_1,m_2,m_3} \overline{\delta x_i^{m_1} \delta y_i^{m_2} \delta z_i^{m_3}}\end{aligned} \tag{3.15}$$

where h can be referred to Eq. (2.13). Equation (3.15) indicates that if the moments $\overline{\delta x_i^{m_1} \delta y_i^{m_2} \delta z_i^{m_3}}$ are known, this integration can be written in an analytical form. In fact, due to the choosing of reconstruction basis in Eq. (2.11), the moments are stored beforehand. To be specific, A_{m_1,m_2,m_3} in Eq. (3.15) can be obtained analytically with $\bar{\Omega}_i$, the coefficients v_k^l of $v_k(\mathbf{x})$ and the moments $\overline{\delta x^{m_1} \delta y^{m_2} \delta z^{m_3}}$ which are stored in the k-exactness reconstruction. Take the third order reconstruction in three-dimension for example, this smooth indicator can be computed as follows

$$\begin{aligned}\frac{\beta_k}{\Omega_i/h} &= a_1^2 + a_2^2 + a_3^2 + h^2(4a_4^2 + a_5^2 + a_6^2 + 4a_7^2 + a_8^2 + 4a_9^2) \\ &\quad + \overline{\delta x_i^2}\left(4a_4^2 + a_5^2 + a_6^2\right) + \overline{\delta x_i \delta y_i}\,(4a_4 a_5 + 4a_5 a_7 + a_6 a_8) \\ &\quad + \overline{\delta x_i \delta z_i}\,(4a_4 a_6 + a_5 a_8 + 4a_6 a_9) + \overline{\delta y_i^2}\left(a_5^2 + 4a_7^2 + a_8^2\right) \\ &\quad + \overline{\delta y_i \delta z_i}\,(a_5 a_6 + 4a_7 a_8 + 4a_8 a_9) + \overline{\delta z_i^2}\left(a_6^2 + a_8^2 + 4a_9^2\right)\end{aligned}$$

where a_l represents v_k^l. These expressions can be easily derived using the Mathematica software. One big advantage of this quadrature-free method is that it is independent of control volume shape, i.e., it can be applied on arbitrary grids, while the traditional quadrature-free way are limited to perform on triangular or tetrahedral grids [15]. For the third order scheme, no more storage is needed and the smooth indicator is easy to compute. For the high order schemes, fourth order at most in this chapter, extra memory is needed for storing the coefficients and the expression for smooth indicator will become more complex. E.g., the smooth indicator for fourth order WENO can be computed as

$$
\begin{aligned}
\frac{\beta}{\Omega_i/h} = {} & a_1^2 + a_2^2 + a_3^2 + h^2(4a_4^2 + a_5^2 + a_6^2 + 4a_7^2 + a_8^2 + 4a_9^2) \\
& + 4h^4(9a_{10}^2 + a_{11}^2 + a_{12}^2 + a_{13}^2 + \tfrac{1}{4}a_{14}^2 + a_{15}^2 + 9a_{16}^2 + a_{17}^2 + a_{18}^2 + 9a_{19}^2) \\
& + \overline{\delta x_i^2}\,(6a_1a_{10} + 2a_{11}a_2 + 2a_{12}a_3 + 4a_4^2 + a_5^2 + a_6^2 + 4h^2(9a_{10}^2 + a_{11}^2 \\
& \qquad + a_{12}^2 + a_{13}^2 + \tfrac{1}{4}a_{14}^2 + a_{15}^2)) \\
& + \overline{\delta x_i \delta y_i} \begin{pmatrix} 2(2a_1a_{11} + 2a_{13}a_2 + a_{14}a_3 + 2a_4a_5 + 2a_5a_7 + a_6a_8) \\ +4h^2(6a_{10}a_{11} + 2a_{11}a_{13} + a_{12}a_{14} + 6a_{13}a_{16} + a_{14}a_{17} \\ +2a_{15}a_{18}) \end{pmatrix} \\
& + \overline{\delta x_i \delta z_i} \begin{pmatrix} 2(2a_1a_{12} + a_{14}a_2 + 2a_{15}a_3 + 2a_4a_6 + a_5a_8 + 2a_6a_9) \\ +4h^2(6a_{10}a_{12} + 2a_{12}a_{15} + 2a_{13}a_{17} + a_{14}a_{11} + a_{14}a_{18} \\ +6a_{15}a_{19}) \end{pmatrix} \\
& + \overline{\delta y_i^2}\,(2a_1a_{13} + 6a_{16}a_2 + 2a_{17}a_3 + a_5^2 + 4a_7^2 + a_8^2 + 4h^2(a_{11} + a_{13}^2 \\
& \qquad + \tfrac{1}{4}a_{14}^2 + 9a_{16}^2 + a_{17}^2 + a_{18}^2)) \\
& + \overline{\delta y_i \delta z_i} \begin{pmatrix} 2(a_1a_{14} + 2a_{17}a_2 + 2a_{18}a_3 + a_5a_6 + 2a_7a_8 + 2a_8a_9) \\ +4h^2(2a_{11}a_{12} + a_{13}a_{14} + a_{14}a_{15} + 6a_{16}a_{17} + 2a_{17}a_{18} \\ +6a_{18}a_{19}) \end{pmatrix} \\
& + \overline{\delta z_i^2}\,(2a_1a_{15} + 2a_{18}a_2 + 6a_{19}a_3 + a_6^2 + a_8^2 + 4a_9^2 + 4h^2(a_{12}^2 + \tfrac{1}{4}a_{14}^2 \\
& \qquad + a_{15}^2 + a_{17}^2 + a_{18}^2 + 9a_{19}^2)) \\
& + \overline{\delta x_i^3} \cdot 2\,(6a_{10}a_4 + a_{11}a_5 + a_{12}a_6) \\
& + \overline{\delta x_i^2 \delta y_i} \cdot 2\,(3a_{10}a_5 + 2a_{13}a_5 + a_{14}a_6 + 2a_{11}(2a_4 + a_7) + a_{12}a_8) \\
& + \overline{\delta x_i^2 \delta z_i} \cdot 2\,(a_{14}a_5 + 3a_{10}a_6 + 2a_{15}a_6 + a_{11}a_8 + 2a_{12}(2a_4 + a_9)) \\
& + \overline{\delta x_i \delta y_i^2} \cdot 2\,(2a_{11}a_5 + 3a_{16}a_5 + a_{17}a_6 + 2a_{13}(a_4 + 2a_7) + a_{14}a_8) \\
& + \overline{\delta x_i \delta y_i \delta z_i} \cdot 4\,(a_{12}a_5 + a_{17}a_5 + a_{11}a_6 + a_{18}a_6 + a_{13}a_8 + a_{15}a_8 \\
& \qquad + a_{14}(a_4 + a_7 + a_9)) \\
& + \overline{\delta x_i \delta z_i^2} \cdot 2\,(a_{18}a_5 + 2a_{12}a_6 + 3a_{19}a_6 + a_{14}a_8 + 2a_{15}(a_4 + 2a_9)) \\
& + \overline{\delta y_i^3} \cdot 2\,(a_{13}a_5 + 6a_{16}a_7 + a_{17}a_8) \\
& + \overline{\delta y_i^2 \delta z_i} \cdot 2\,(a_{14}a_5 + a_{13}a_6 + 4a_{17}a_7 + 3a_{16}a_8 + 2a_{18}a_8 + 2a_{17}a_9) \\
& + \overline{\delta y_i \delta z_i^2} \cdot 2\,(a_{15}a_5 + a_{14}a_6 + 2a_{18}a_7 + 2a_{17}a_8 + 3a_{19}a_8 + 4a_{18}a_9) \\
& + \overline{\delta z_i^3} \cdot 2\,(a_{15}a_6 + a_{18}a_8 + 6a_{19}a_9) + \text{high order moment terms.}
\end{aligned}
$$

To avoid the extra storage, the terms corresponding to the high-order moments which are not stored in the basis are cut off, e.g., for the fourth order scheme in three dimension, $A_{m_1,m_2,m_3}\overline{\delta x_i^{m_1}\delta y_i^{m_2}\delta z_i^{m_3}}$, $m_1 + m_2 + m_3 = 4$ are abandoned in the evaluation of smooth indicator. Numerical tests in Sects. 3.8 and 3.9 will validate the efficiency of this simplification.

In terms of these smoothness indicators, the weight of each candidate reconstruction is determined by

$$w_k = \frac{\alpha_k}{\sum_{k=0}^{Ns}\alpha_k}, \quad \alpha_k = \frac{\lambda_k}{(\varepsilon + \beta_k)^p}$$

Here, we choose $p = 4$. ε is a small positive number to avoid possible division by zero which is chosen to be 10^{-8} in this chapter, and λ_k is the linear weight. The linear weights can be chosen as arbitrary positive numbers whenever the scheme is stable. However, following the same reasoning of [14], it is beneficial to emphasis the contribution of the primary reconstruction (PR). Therefore, λ_k is computed by

$$\lambda_k = \begin{cases} \lambda_0, & k = 0 \\ 1, & \text{else} \end{cases}$$

λ_0 can be chosen as 100 and 10,000 in our computation. In Sect. 3.8, we will give numerical examples on the influence of this parameter. Without specification, λ_0 is set to be 100 which is favored for strong shock capturing. The final polynomial for the cell Ω_i would be the weighted summation of all candidate polynomials,

$$\tilde{v}(\mathbf{x}) = \sum_{k=0}^{Ns} w_k \cdot v_{i,k}(\mathbf{x})$$

which is used to compute the right and left state in the flux.

There are a number of potential advantages of the k-exact WENO scheme compared to other high-order FV schemes. (1) The present FV scheme is principally similar to the WENO-1 schemes. Therefore, the present scheme can take advantage of the flexibility offered by this type of WENO schemes, for example, weighting the central stencil with a large linear weight. Specifically, the primary k-exact reconstruction of the present FV scheme is basically the same as the reconstruction on central stencil of WENO-1. The main difference between the present FV scheme and WENO-1 is that other candidate reconstructions for WENO-1 are replaced by the secondary reconstructions which can be deduced easily from the primary k-exact reconstructions of neighboring cells without the need of stencil identification and solution of the linear system of equations. Therefore, the present FV scheme is more efficient than the WENO-1 schemes of the same accuracy. (2) When comparing to the WENO-2 schemes, the primary k-exact reconstruction of the present scheme is comparable to the reconstruction on big stencil of WENO-2 with the same accuracy in terms of stencil and computational cost (if the inversion of the reconstruction matrix is precomputed and stored). The secondary reconstructions are functionally similar to the reconstructions on small stencils but do not relay on any stencil identification technique. Therefore, the design and implementation of the present FV scheme are more straightforward and systematic. Another significant advantage of the present scheme is that it automatically removes the problem of negative linear weights of WENO-2, which makes the present scheme more robust especially for schemes higher than third order. (3) The present scheme has an excellent shock capturing capability when comparing to the traditional k-exact reconstruction approach. Similar to other WENO schemes, the present FV scheme based on primary and secondary k-exact reconstructions is uniformly accurate for smooth flow fields and captures the shock wave and contact discontinuities essentially free from numerical oscillations. When the shock detecting technique presented in the next subsection is adopted, the present scheme is only slightly expensive than the unlimited k-exact approach in terms of the computational cost, but is equipped with excellent shock capturing capability.

3.4 The WBAP Limiters

In the last section, we use the secondary reconstruction and the k-exact WENO limiters to suppress the numerical oscillations. Because of the large stencils in the limiting and the complicated formulas to compute the WENO smooth indicator, the k-exact WENO schemes take more time than the traditional slope limiters. In this section, an new accuracy preserving limiting procedure for the high-order FV schemes will be proposed to raise the efficiency. Before the detailed introduction of the procedure, some general considerations will be presented first.

(1) Accuracy preserving. In this section, the limiting procedure is also applied to several candidate reconstructions of the same order of accuracy. On cell Ω_i, the k-exact reconstruction presented in Sect. 2.3 is used as the first candidate reconstruction and is called the PR hereafter. Other candidate reconstructions are constructed by means of the secondary reconstruction or SR for short. The SR has been proposed in Sect. 3.2 from a different perspective to show its construction better.
(2) Compactness. From the point of view of practical implementation, the limiting procedure should be as compact as possible. In this chapter, the implementation of the limiter only uses the information of current cell and its direct face neighbors. The compactness of the present procedure relies on the compactness of the SR. The SR is a special continuation of the PR on a neighboring cell. Referring to Fig. 2.2, a SR for Ω_i can be derived from the PR on Ω_{1a}, which is denoted with $u_{1a\to i}(\mathbf{x})$. This procedure is very compact using only the PR on Ω_{1a}. In the present limiting procedure, only 3 SRs are needed which are derived from cells Ω_{1a}, Ω_{2a} and Ω_{3a} in Fig. 2.2. Therefore, the SRs for Ω_i can be written in the following general forms

$$u_{j\to i}(\mathbf{x}) = \bar{u}_i + \sum_{l=1}^{DOF} u^l_{j\to i}\varphi_{l,i},\ j \in \{1a, 2a, 3a\}. \tag{3.16}$$

(3) Robustness. For high order FV schemes, a straightforward generalization of the second order limiting strategy is to apply the limiting function to the coefficients or scaled derivatives of the PR described by Eq. (2.10), and the SRs presented in Eq. (3.16), which yields

$$\tilde{u}_i(\mathbf{x}) = \bar{u}_i + \sum_{l=1}^{DOF} L(u^l_i, u^l_{1a\to i}, u^l_{2a\to i}, u^l_{3a\to i})\varphi_{l,i}, \tag{3.17}$$

where $\tilde{u}_i(\mathbf{x})$ is the reconstruction polynomial after limiting, and L is the limiting function corresponding to WBAP whose specific form for high-order FV schemes will be presented in Sect. 3.4.1.

3.4.1 The Introduction of WBAP Functions

In this section, the limiters based on the WBAP will be presented. For the purpose of comparison, the limiting procedures of TVD-MUSCL and BAP will also be briefly introduced. For simplicity, we consider the initial value problem of the scalar conservation law,

$$\begin{aligned} u_t &= -f(u)_x \\ u(x, t = 0) &= u_0(x) \end{aligned} \tag{3.18}$$

where $u_0(x)$ is either a piecewise smooth function with compact support or a periodic function. On a uniform grid, integrating Eq. (3.18) over the control volume $C_i = [x_{i-\frac{1}{2}}, x_{i+\frac{1}{2}}]$, we obtain the standard semi-discrete finite volume scheme

$$\frac{d}{dt}\bar{u}_i = \frac{1}{\Delta x}\left[F(u^-_{i-1/2}, u^+_{i-1/2}) - F(u^-_{i+1/2}, u^+_{i+1/2})\right] \tag{3.19}$$

where $\bar{u}_i$ is the cell average of u over C_i, $F(u^-_{i-1/2}, u^+_{i-1/2})$ is the numerical flux computed by a certain Riemann solver or flux splitting technique, and the left and right states in Eq. (3.19) are evaluated by applying the reconstruction technique. Here we only consider the spatially second order finite volume scheme for which the linear reconstruction on C_i yields

$$u_{i-1/2}^{+} = \bar{u}_i - \frac{\Delta x}{2}\sigma_i^L$$
$$u_{i+1/2}^{-} = \bar{u}_i + \frac{\Delta x}{2}\sigma_i^R \tag{3.20}$$

where σ_i^L and σ_i^R are the reconstructed gradients or slopes. In order to suppress the numerical oscillations near the discontinuities, a limiting procedure has to be applied on the slopes.

One of the most successful limiting procedures within the framework of the finite volume methods is the TVD-MUSCL approach, in which the limited slope is computed by

$$\sigma_i^L = \frac{\delta_{i-1/2}}{\Delta x}\varphi(\theta_i)$$
$$\sigma_i^R = \frac{\delta_{i+1/2}}{\Delta x}\varphi(\frac{1}{\theta_i}) \tag{3.21}$$

where $\delta_{i+1/2} = \bar{u}_{i+1} - \bar{u}_i$, $\theta_i = \frac{\delta_{i+1/2}}{\delta_{i-1/2}}$, and $\varphi(\theta_i)$is the limiter of the TVD scheme. Sweby [17] defined a class of limiters for the second order TVD schemes satisfying

$$\max(0, \min(\theta_i, 1)) \leq \varphi(\theta_i) \leq \max(0, \min(2\theta_i, 1), \min(\theta_i, 2)). \tag{3.22}$$

Note that $\varphi(\theta_i)$ is usually asked to have a symmetry of

$$\frac{\varphi(\theta_i)}{\theta_i} = \varphi(\frac{1}{\theta_i})$$

so that the slope is uniquely determined independent of the direction of upwinding, i.e.,

$$\sigma_i^L = \sigma_i^R.$$

The TVD-MUSCL discretization captures the discontinuities in very high resolution but is difficult to be applied directly on unstructured grids.

The BAP of Choi and Liu [18] is a limiting procedure that can suppress the oscillation near the discontinuity and can be applied on unstructured grids. Using the BAP, the limited slopes in (3.20) can be written as

$$\sigma_i^L = \sigma_i^R = \frac{1}{\Delta x}L(\delta_{i+1/2}, \delta_{i-1/2}), \tag{3.23}$$

where L is defined by

$$L(a_0, a_1) = B^{-1}\left[\frac{B(a_0) + B(a_1)}{2}\right]. \tag{3.24}$$

The function B is called the biased function. There are many possible biased functions, and one example is

$$B(x) = \frac{x}{\sqrt{1+x^2}}, \; B^{-1}(x) = \frac{x}{\sqrt{1-x^2}}.$$

In [18], this procedure performed very well when used in conjunction with the reconstruction of the upwind fluxes obtained by some flux splitting techniques. However, the numerical test cases in [19] show that it is sometimes not sufficient to control the oscillations near the discontinuities when used in the MUSCL type finite volume schemes. Moreover, its mechanism to suppress the numerical oscillations is not well understood.

To overcome these deficiencies, a new limiting procedure, namely the WBAP, will be proposed in this section. In the context of one-dimensional finite volume scheme, we present two versions of the WBAP:

Version1:

$$\begin{aligned} \sigma^L &= L\left(\frac{\delta_{i-1/2}}{\Delta x}, \frac{\delta_{i+1/2}}{\Delta x}\right) \\ \sigma^R &= L\left(\frac{\delta_{i+1/2}}{\Delta x}, \frac{\delta_{i-1/2}}{\Delta x}\right), \end{aligned} \tag{3.25}$$

Version2:

$$\sigma^L = \sigma^R = L\left(\bar{\sigma}_i, \bar{\sigma}_{i+1}, \bar{\sigma}_{i-1}\right) \tag{3.26}$$

where $\bar{\sigma}_i = \frac{1}{2\Delta x}(\delta_{i+1/2} + \delta_{i-1/2})$.

In Eqs. (3.25) and (3.26), L is the limiting function of WBAP. Before detailing the form of L, the difference between Version 1 and Version 2 will be discussed first. Version 1 is similar in form to the TVD-MUSCL approach (Eq. (3.21)) and the BAP (Eq. (3.23)) in which the left and right gradients (or slopes) are computed and used as the input of the limiting function. Using this approach, it is possible to study the behavior of L by comparing with the limiters of the TVD-MUSCL approach. However, Version 1 is difficult to be applied directly on unstructured grids and does not satisfy the condition of symmetry. In Version 2, the gradients are computed by central difference, which is identical to the gradients computed by the least-squares method on unstructured grids. Moreover, the arguments of L in Eq. (3.26) are the gradients of current cell and its immediate neighbors, which make it possible to generalize Version 2 WBAP to unstructured grids as long as the function L accepts multiple arguments.

In this chapter, L is in the following form

$$L\left(a_0, a_1, a_2, \ldots, a_J\right) = a_0 \cdot W\left(1, \frac{a_1}{a_0}, \frac{a_2}{a_0}, \ldots, \frac{a_J}{a_0}\right) \tag{3.27}$$

where a_0 is the first argument of L. The function W is called the WBAP limiter which is functionally identical to the limiters of the TVD-MUSCL schemes and is defined by

$$W(\theta_0, \theta_1, \theta_2, \ldots, \theta_J) = B^{-1}\left[\sum_{j=0}^{J} \omega_j B(\theta_j)\right] \tag{3.28}$$

where $\theta_j = a_j/a_0$ ($\theta_0 = 1$), B is the biased function and ω_j, $j = 1, \ldots, J$ are the weights. In this chapter, we consider the following biased function and weight function which are, respectively

$$B(x) = \frac{x}{\sqrt{\varepsilon^2 + x^2}}, \quad B^{-1}(x) = \frac{\varepsilon x}{\sqrt{1 - x^2}} \tag{3.29}$$

and

$$\omega_j = \frac{\alpha_j}{\sum_{j=0}^{J} \alpha_j}, \quad \alpha_j = \frac{\lambda_j}{B(\theta_j)^2 + \delta}. \tag{3.30}$$

In Eq. (3.29), ε is a free parameter whose value range will be discussed in the next section. In Eq. (3.30), δ is a small positive number to avoid possible division by zero, which is chosen as $\delta = 10^{-10}$, and λ_j is determined by

$$\lambda_j = \begin{cases} n, & \text{if } j = 0 \\ 1, & \text{else} \end{cases}$$

If $n = 1$, the weights in Eq. (3.30) are solely determined by the smoothness indicator. If $n \geq 1$, the contribution of the central cell is emphasized. At present, it is worthwhile to introduce some simple properties of WBAP in terms of the following remarks.

Remark 1 Because of the specific form of L in Eq. (3.27), the WBAP is self-similar, i.e., it is invariant when the spatial and time variables are scaled by the same factor. Furthermore, the function W acts exactly as the limiters of the TVD-MUSCL approach.

Remark 2 Before applying the inverse of the biased function in Eq. (3.28), the $B(a_j)$ are weighted averaged. The biased functions of WBAP are also different with those of BAP. By the introduction of the parameter ε, the asymptotic behavior of WBAP can be studied to reveal its mechanism for suppressing the numerical oscillations, which will be discussed in the next section.

Remark 3 The WBAP is continuously differentiable. The nondifferentiable limiters are sensitive to small perturbations and may cause slow convergence, while the differentiable one enables a smooth transition from the high-order solution in smooth region to the first order solution in the vicinity of discontinuity.

Remark 4 When the solution is smooth, it is easy to prove that

$$u^{\pm}_{i+1/2} = u^{\text{exact}}_{i+1/2} + e(x_{i+1/2})\Delta x^2 + O(\Delta x^3) \tag{3.31}$$

and $e(x)$ is Lipschitz continuous. Following the analysis of Harten [2], Eq. (3.31) guarantees the second order accuracy of the scheme. At the local extrema $e(x)$ is not Lipschitz continuous and the scheme will reduce to first order in term of L_∞ norm. We note that a simple technique presented in [20] can be used to overcome this drawback. Furthermore, when the limiter based on WBAP is used in r-th order reconstruction, the resulting scheme will be r-th order except near the roots of higher order derivative of u where the scheme may locally reduce to $(r-1)$-th order accurate in the L_∞ norm. This property will be further studied in a future paper where high-order reconstruction will be used.

3.4.2 The Mechanism for Controlling the Numerical Oscillations

In this section, the mechanism of WBAP for suppressing the numerical oscillations is discussed. We first consider the Version 1 of WBAP. In this case, W in Eq. (3.25) behaves exactly the same as the limiter in the TVD-MUSCL approach.

In the WBAP limiter, ε is used to adjust the dissipation of the limiter. We note that as $\varepsilon \to 0$, the limit of $W(1, \theta_i)$ is

$$\lim_{\varepsilon \to 0} W(1, \theta_i) = \max\left(0, \frac{\sqrt{(1+n)\theta_i}}{\sqrt{1+n\theta_i^2}}\right) \tag{3.32}$$

for the biased function presented in Eq. (3.29). It is easy to prove that when $n \le 3$, $\lim_{\varepsilon \to 0} W(1, \theta_i)$lies in the TVD region of Sweby with the superbee and minmod limiters as its upper and lower bounds (Eq. (3.22)). Therefore, for some values of n, $W(1, \theta_i)$ is a limiter with asymptotically TVD property. This behavior is shown in Fig. 3.2. In this figure, it is also shown that even if ε is chosen as a finite value, for example $\varepsilon = 1$, $W(1, \theta_i)$ is also inside or almost inside the TVD region when $\theta_i \ge 0$. Furthermore, for a finite value of ε, $W(1, \theta_i)$ is always smooth. The asymptotically TVD behavior of $W(1, \theta_i)$ explains why the WBAP is capable of removing the numerical oscillations near the discontinuities at least for the one-dimensional case.

For the Version 2 of WBAP, one of its remarkable advantages is that it can be straightforwardly applied to multidimensional unstructured grid cases. Therefore, its shock capturing mechanism will be studied in more detail. We note that the asymptotic behavior of the Version 2 WBAP can be also studied. To be general, we consider Eq. (3.28) with arbitrary number of arguments. When $\varepsilon \to 0$, we have

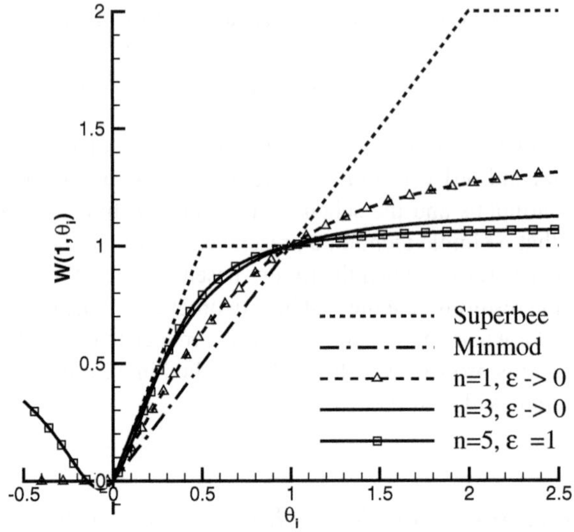

Fig. 3.2 The behavior of the WBAP. Reprinted from Ref. [21], Copyright 2011, with permission from Elsevier

$$W^{L1}(1,\theta_1,\ldots,\theta_J)=\lim_{\varepsilon\to 0}W(1,\theta_1,\ldots,\theta_J)=\begin{cases}\sqrt{\dfrac{n+J}{n+\sum\limits_{k=1}^{J}1/\theta_k^2}}, & \text{if } \theta_1,\ldots,\theta_J>0\\ 0, & \text{otherwise.}\end{cases} \tag{3.33}$$

Equation (3.33) is a new limiter denoted as WBAP-L1 hereafter which can be used to replace W in Eq. (3.27). Comparing with Eq. (3.32), WBAP-L1 can be considered as a logical generalization of the TVD limiter to multiple arguments cases. The Version 2 WBAP provides new TVD-like limiters that can be applied to multidimensional problems. And furthermore, in what follows, we will prove that for one-dimensional case, WBAP-L1 satisfies a certain MP.

Lemma 3.1 *Assume WBAP-L1 limiter is used in the limiting procedure of Eq.* (3.26) *and Eq.* (3.27). *If* $n \le 14$, *then*

$$u^+_{i-1/2}, u^-_{i+1/2} \in [m, M],$$

where m and M are the minimum and maximum values of $\{\bar{u}^n_{i-2}, \bar{u}^n_{i-1}, \bar{u}^n_i, \bar{u}^n_{i+1}, \bar{u}^n_{i+2}\}$.

The proof can be seen in Appendix B.

Theorem 3.1 *For the discrete finite volume scheme Eq.* (3.19) *with a monotone Lipschitz continuous flux function, if Eq.* (3.33) *is used in the limiting procedure of Eqs.* (3.26) *and* (3.27) *and* $n \le 14$, *then the scheme ensures the MP, i.e.,*

$$\bar{u}^{n+1}_i \in [m, M]$$

under the CFL condition, where m and M are the minimum and maximum values of $[\bar{u}_{i-3}^n, \bar{u}_{i-2}^n, \bar{u}_{i-1}^n, \bar{u}_i^n, \bar{u}_{i+1}^n, \bar{u}_{i+2}^n, \bar{u}_{i+3}^n]$.

Proof Using Lemma 1, one can obtain

$$u_{i-1/2}^-, u_{i-1/2}^+, u_{i+1/2}^-, u_{i+1/2}^+ \in [m, M].$$

The application of the Theorem 2.2 of [4] to the case of linear reconstruction leads to $\bar{u}_i^{n+1} \in [m, M]$ under the CFL condition.

The theorem of this paper shows that the WBAP limiter asymptotically satisfies the MP when $\varepsilon \to 0$. This property ensures the shock capturing capability of the present limiter. Similar to the MLP [7], the MP satisfied by the WBAP-L1 involve a larger stencil than the MP corresponding to the Barth limiter [5]. Therefore, the present limiter is less restrictive in the admissible range of the limited gradient.

Next, the influence of ε in Eq. (3.29) will be studied. We have already shown that, when $\varepsilon \to 0$, the resulting WBAP-L1 limiter satisfies the MP. It is interesting to note that when $\varepsilon \to \infty$, the WBAP also has a limit in the following form,

$$W^{L2}(1, \theta_1, \ldots, \theta_J) = \lim_{\varepsilon \to +\infty} W(1, \theta_1, \ldots, \theta_J) = \frac{n + \sum\limits_{k=1}^{J} 1/\theta_k}{n + \sum\limits_{k=1}^{J} 1/\theta_k^2} \tag{3.34}$$

which is termed as the WBAP-L2 limiter. Although it has not been proved that W^{L2} satisfies the MP, extensive numerical tests in Sect. 3.9 show that W^{L2} can also produce essentially oscillation-free numerical results. In fact, if the WBAP-L2 limiter is modified as

$$W^{L2}(1, \theta_1, \ldots, \theta_J) = \begin{cases} \dfrac{n + \sum\limits_{k=1}^{J} 1/\theta_k}{n + \sum\limits_{k=1}^{J} 1/\theta_k^2}, & \text{if } \theta_1, \ldots, \theta_J > 0 \\ 0, & \text{otherwise,} \end{cases}$$

the modified WBAP-L2 limiter is also MP-satisfying. Furthermore, for the case of $\theta_1, \ldots, \theta_J \geq 0$, we can prove that

$$W^{L1}(1, \theta_1, \ldots, \theta_J) \geq W^{L2}(1, \theta_1, \ldots, \theta_J) \tag{3.35}$$

by using the Cauchy-Schwarz inequality, which means that W^{L2} is in fact more dissipative than W^{L1} for the monotone numerical solutions. We consider a special case $\theta = \theta_1 = \cdots = \theta_J$, the influence of ε can be shown clearly in Fig. 3.3 in terms of the $W - \theta$ curves. According to this figure, the WBAP-L1 limiter is the least dissipative while the WBAP-L2 limiter is the most dissipative for $\theta \geq 0$. For other ε, the dissipation of WBAP lies basically in-between WBAP-L1 and WBAP-L2.

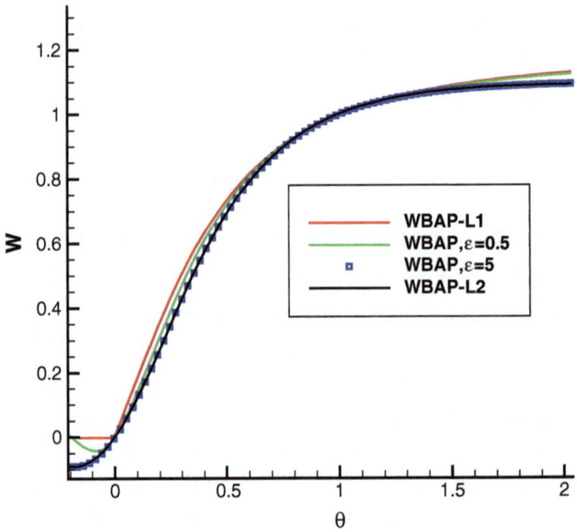

Fig. 3.3 The influence of ε in WBAP, $n = 5$. Reprinted from Ref. [21], Copyright 2011, with permission from Elsevier

The results suggest that ε can be chosen arbitrary from infinite small to infinite large, and a larger value of ε will result in a more dissipative WBAP limiter. Because the two limiters, WBAP-L1 and WBAP-L2 are computationally more efficient than WBAP, they will be applied in most of the test cases of the present paper.

3.4.3 The Successive Limiting Procedure

Based on the SRs and the limiters presented in Sects. 3.1 and 3.4.1, the limiting procedure outlined at the head of this section can now be summarized as follows:

The limiting procedure-1 (an intuitive but unsatisfactory approach)

After performing the PR and SRs, the following candidate reconstructions are obtained

$$\begin{aligned} u_i(\mathbf{x}) &= \bar{u}_i + \sum_{l=1}^{DOF} u_i^l \varphi_{l,i}(\mathbf{x}), \\ u_{j\to i}(\mathbf{x}) &= \bar{u}_i + \sum_{l=1}^{DOF} u_{j\to i}^l \varphi_{l,i}(\mathbf{x}),\ j \in \{1a, 2a, 3a\}. \end{aligned}$$

The limiting procedure of Eq. (3.34) is applied to the coefficients of the PR and SRs of Ω_i, and the limited coefficient $\tilde{u}_i^l$ is

$$\tilde{u}_i^l = L(u_i^l, u_{1a\to i}^l, u_{2a\to i}^l, u_{3a\to i}^l).$$

Using the limited coefficients the reconstruction polynomial after limiting is

$$\tilde{u}_i(\mathbf{x}) = \bar{u}_i + \sum_{l=1}^{DOF} \tilde{u}_i^l \varphi_{l,i}.$$

The limiting procedure-1 is very simple and efficient. However, its performance is not satisfactory. Figures 3.4 and 3.5 show the results of this limiting procedure in solving one-dimensional scalar linear advection equations

$$\frac{\partial u}{\partial t} + \frac{\partial u}{\partial x} = 0$$

using the sine wave and square wave as the initial conditions. The grid number is 100 and the computation is carried out until the waves travel twice of the domain length using the periodic boundary condition. The WBAP-L2 limiter is adopted in the simulation and n in Eq. (3.34) is chosen as $n = 1$ and $n = 100$, respectively. The case with a sine wave initial condition is shown in Fig. 3.4. Although the numerical solution of u is very close to the analytical one for both $n = 1$ and $n = 100$, the solution of u_x and u_{xx} show different behaviors. Both u_x and u_{xx} can be predicted in high accuracy at $n = 100$, while they are predicted with much larger errors at $n = 1$. For the case with a square wave initial condition, Fig. 3.5 indicates that when $n = 1$, the solution is very dissipative; and when $n = 100$, numerical oscillations can be observed near the discontinuities. The results in Figs. 3.4 and 3.5 suggest that this limiting procedure is very sensitive to n. Furthermore, the requirement for n is contradictory for smooth solution and solution with discontinuities. For the case of smooth solution, a larger value of n is required to enhance the accuracy; and for the case of discontinuous solution, a smaller value of n is beneficial in reducing the numerical oscillations. This observation makes it difficult to determine an optimal value of n in practical simulation.

To overcome these deficiencies, a successive limiting procedure is proposed in this section. For the fourth order FV schemes, the successive limiting procedure denoted by limiting procedure-2 is presented below. For the second and third order schemes, the procedures are similar.

The limiting procedure-2 (successive limiting)

(I) On cell Ω_i, the third order-scaled derivatives of the SRs are computed according to Eq. (3.13), namely

$$\hat{u}^l_{j\to i} = u^l_j, l = 6, 7, 8, 9, j \in \{1a, 2a, 3a\}$$

The limited third order-scaled derivatives on Ω_i are computed by

$$\tilde{u}_i^l = L\left(u_i^l, \hat{u}^l_{1a\to i}, \hat{u}^l_{2a\to i}, \hat{u}^l_{3a\to i}\right), l = 6, 7, 8, 9.$$

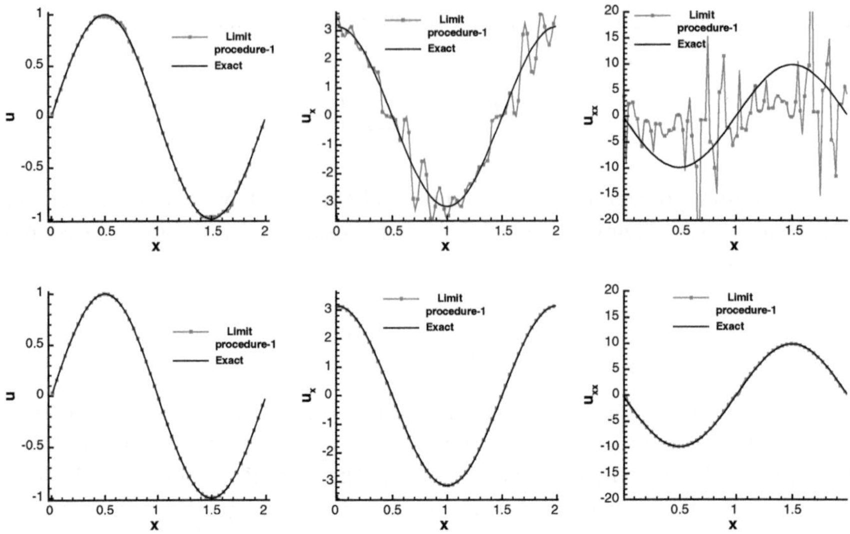

Fig. 3.4 Solution of limiting procedure-1 in solving scalar equation with sine wave initial condition. Above,$n = 1$; below,$n = 100$. From left to right are different scaled derivatives, u, u_x, u_{xx}, respectively. Reprinted from Ref. [22], Copyright 2012, with permission from Elsevier

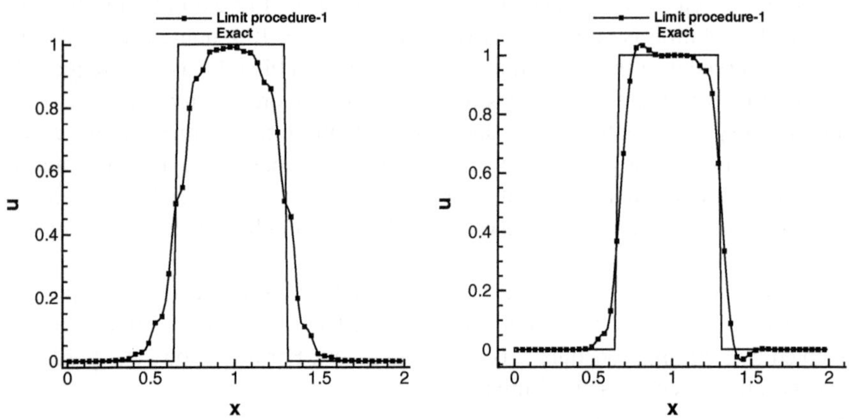

Fig. 3.5 Solution of limiting procedure-1 in solving scalar equation with square wave initial condition. Left,$n = 1$; right,$n = 100$. Reprinted from Ref. [22], Copyright 2012, with permission from Elsevier

This operation is carried out on all control volumes and the $\tilde{u}_i^l (l = 6, 7, 8, 9)$ are stored.

(II) The second order-scaled derivatives of the SRs are computed using the limited third order- scaled derivatives through

$$\begin{aligned}
\hat{u}^3_{j\to i} &= u^3_j + 3\tilde{u}^6_j \Delta x + \tilde{u}^7_j \Delta y,\\
\hat{u}^4_{j\to i} &= u^4_j + 2\tilde{u}^7_j \Delta x + 2\tilde{u}^8_j \Delta y,\\
\hat{u}^5_{j\to i} &= u^5_j + \tilde{u}^8_j \Delta x + 3\tilde{u}^9_j \Delta y,\\
\Delta x &= x_i - x_j,\ \Delta y = y_i - y_j,\ j \in \{1a, 2a, 3a\}.
\end{aligned}$$

The secondary order-scaled derivatives are limited using

$$\tilde{u}^l_i = L\left(u^l_i, \hat{u}^l_{1a\to i}, \hat{u}^l_{2a\to i}, \hat{u}^l_{3a\to i}\right), l = 3, 4, 5.$$

Again this operation is carried out over all control volumes and $\tilde{u}^l_i (l = 3, 4, 5)$ are stored.

(III) The first order-scaled derivatives of the SRs are computed using the limited second order and third order-scaled derivatives through

$$\begin{aligned}
\hat{u}^1_{j\to i} &= u^1_j + 2\tilde{u}^3_j \Delta x + \tilde{u}^4_j \Delta y + 3\tilde{u}^6_j \Delta x^2 + 2\tilde{u}^7_j \Delta x \Delta y + \tilde{u}^8_j \Delta y^2,\\
\hat{u}^2_{j\to i} &= u^2_j + \tilde{u}^4_j \Delta x + 2\tilde{u}^5_j \Delta y + \tilde{u}^7_j \Delta x^2 + 2\tilde{u}^8_j \Delta x \Delta y + 3\tilde{u}^9_j \Delta y^2,\\
\Delta x &= x_i - x_j,\ \Delta y = y_i - y_j,\ j \in \{1a, 2a, 3a\}.
\end{aligned}$$

Then the first order-scaled derivatives are limited using

$$\tilde{u}^l_i = L\left(u^l_i, \hat{u}^l_{1a\to i}, \hat{u}^l_{2a\to i}, \hat{u}^l_{3a\to i}\right), l = 1, 2.$$

After these three steps, all scaled derivatives are limited and the reconstruction polynomial after limiting is obtained.

The key ingredient of the successive limiting procedure is to apply the limiter to high-order-scaled derivatives first over the entire control volumes of the domain and the lower order-scaled derivatives of the SRs are evaluated in terms of the already limited high-order-scaled derivatives. This approach yields a very robust limiting procedure. Now we apply the successive limiting procedure to the same test cases mentioned previously in this section and the results are shown in Figs. 3.6 and 3.7, which are corresponding to sine wave and square wave cases, respectively. It can be observed that for the sine wave case, the results become less sensitive to the value of n and the accuracy is improved also. For the square wave case, the discontinuities can be captured in higher resolution and no oscillations are observed even for $n = 100$. Although a rigorous analysis of this behavior is difficult, a tentative explanation can be given here. (1) In numerical simulation, higher order derivatives are usually predicted with larger errors. (2) In the limiting procedure-1, during the SRs, the higher order derivatives are used to compute the lower order derivatives according to Eq. (3.13). And in this process, the errors of the high order derivatives are transferred to the lower order derivatives and the numerical solutions are thus contaminated. (3) In the limiting procedure-2, i.e., the successive limiting procedure, the lower order derivatives are computed using the limited high-order derivatives. This practice can improve the accuracy and robustness of the numerical schemes mainly due to two

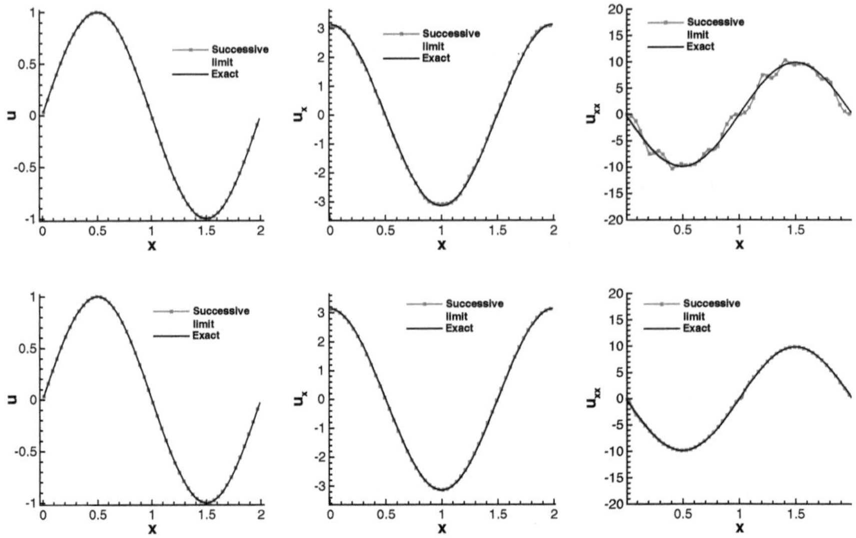

Fig. 3.6 Solution of successive limiting in solving scalar equation with sine wave initial condition. Above, $n = 1$; below, $n = 100$. From *left* to *right* are different scaled derivatives, u, u_x, u_{xx} respectively. Reprinted from Ref. [22], Copyright 2012, with permission from Elsevier

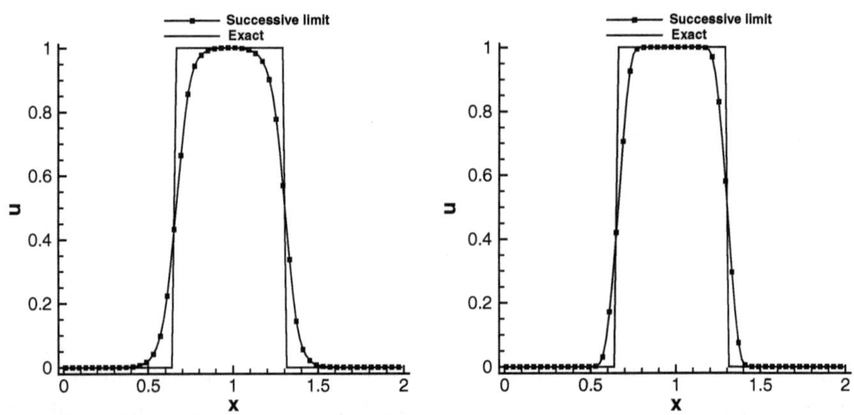

Fig. 3.7 Solution of successive limiting in solving scalar equation with square wave initial condition. *Left*, $n = 1$; *right*, $n = 100$. Reprinted from Ref. [22], Copyright 2012, with permission from Elsevier

reasons. The first, is that the limited high-order derivatives can be actually more accurate than the unlimited ones since the oscillations of the high-order derivatives are suppressed by the limiter. The second one is that the successive limiting procedure blocks transferring the errors of the higher order derivatives to lower order ones.

For system of the hyperbolic conservation laws, the proposed successive limiting procedure can be applied directly to each component of the conservative variables.

The use of conservative variables makes the limiting procedure very efficient and simple for multidimensional problems especially for equations with source terms. When the discontinuities are not very strong, this approach can yield very good results. For very strong discontinuities, a small amount of oscillations can be observed, which is common to all limiting procedures that do not use characteristic variables. To suppress the numerical oscillations more effectively, the limiting procedure can be carried out in terms of the characteristic variables, which will be described below.

For the fourth order FV schemes, the limiting procedure-2 consists of 3 steps. A careful examination reveals that these 3 steps are almost identical except different range of l. Each step can be further divided into 2 substeps. In the first substep, the coefficients of SRs are computed in a successive manner for some l. In the second substep, the limiting function is applied to the coefficients of the PR and SRs to obtain the limited coefficients. For example, for the second step of the limiting procedure-2, the first substep is,

$$\begin{aligned} \hat{u}^3_{j\to i} &= u^3_j + 3\tilde{u}^6_j \Delta x + \tilde{u}^7_j \Delta y, \\ \hat{u}^4_{j\to i} &= u^4_j + 2\tilde{u}^7_j \Delta x + 2\tilde{u}^8_j \Delta y, \\ \hat{u}^5_{j\to i} &= u^5_j + \tilde{u}^8_j \Delta x + 3\tilde{u}^9_j \Delta y, \\ \Delta x &= x_i - x_j, \Delta y = y_i - y_j, j \in \{1a, 2a, 3a\}. \end{aligned} \tag{3.36}$$

and the range of l is $l = 3, 4, 5$. The second substep is the limiting procedure

$$\tilde{u}^l_i = L\left(u^l_i, \hat{u}^l_{1a\to i}, \hat{u}^l_{2a\to i}, \hat{u}^l_{3a\to i}\right), l = 3, 4, 5. \tag{3.37}$$

Therefore, it is sufficient to discuss how to implement the substeps 1 and 2 when the characteristic variables are considered.

For the first substep, it is easy to show it is not necessary to implement Eq. (3.36) in terms of the characteristic variables. When the linearized characteristics are considered, the computation of the characteristic variables is a linear operation, and Eq. (3.36) is also linear. Therefore, the use of the conservative variables or the characteristic variables to evaluate Eq. (3.36) yield exactly the same results. For the second substep, because the limiting function is nonlinear, the use of the conservative variables or the characteristic variables will give different results. According to this discussion, it is only necessary to describe how to implement substep 2 in terms of the characteristic variables.

We notice $\hat{u}^l_{j\to i}$ in Eq. (3.36) is a component of the scaled derivatives of the conservative variables. Therefore, using Eq. (3.36), the vector of the scaled derivatives of the conservative variables $\hat{U}^l_{j\to i}$ can be also evaluated. The scaled derivatives for conservative variables of the PR denoted by U^l_i are already known through the k-exact reconstruction. Therefore, the corresponding characteristic variables can be obtained by using

$$\begin{aligned} V^{l,m}_i &= (R^m)^{-1} U^l_i, \\ \hat{V}^{l,m}_{j\to i} &= (R^m)^{-1} \hat{U}^l_{j\to i}, \end{aligned} \tag{3.38}$$

where R^m is the matrix of right eigenvectors of the Jacobian matrix of the governing equations associated with the direction $\mathbf{n}_m$. For the triangular control volume, $m = 1, 2, 3$. The components of $V_i^{l,m}$ and $\hat{V}_i^{l,m}$ are denoted by $v_i^{l,m}$ and $\hat{v}_i^{l,m}$ respectively. Using the characteristic variables, Eq. (3.37) is replaced by

$$\tilde{v}_i^{l,m} = L\left(v_i^{l,m}, \hat{v}_{1a\to i}^{l,m}, \hat{v}_{2a\to i}^{2,m}, \hat{v}_{3a\to i}^{l,m}\right). \tag{3.39}$$

Equation (3.39) is evaluated 3 times for $m = 1, 2, 3$, and the corresponding components of the conservative variables can be easily deduced which are denoted by $\tilde{u}_i^{l,m}$. Therefore, we obtain 3 sets of limited coefficients. From the point of view of practical applications, it is better to have a unique reconstruction polynomial, which is especially important for the problem with source terms [14]. In the present chapter, $\tilde{u}_i^{l,m}$, $m = 1, 2, 3$ are further limited using

$$\tilde{u}_i^l = L(\tilde{u}_i^{l,1}, \tilde{u}_i^{l,2}, \tilde{u}_i^{l,3})$$

to get a unique set of $\tilde{u}_i^l$, and this WBAP limiter must be used with $n = 1$ as we treat every direction equally. A similar idea has been used by Dumbser et al [15] in their implementation of the WENO schemes in terms of the characteristic variables.

This limiting procedure is very compact in form. In fact, only the cell of interest Ω_i and its immediate neighbors $\{1a, 2a, 3a\}$ are involved in the limiting procedure. This is rarely seen in other accuracy preserving high order limiting procedures. Because the secondary reconstruction is derived from the reconstruction on neighboring cells, the limiting procedure uses a larger stencil in an implicit way, which however does not affect the compactness of the limiting procedure in practical implementation.

3.5 The Group Weighted Limiters

The last section provides a simple yet efficient limiting procedure which is similar to the moment limiter, Hierarchical/pointwise Hierarchical reconstructions. They all limit the coefficients of the reconstructed polynomial from the highest order to the lowest one and only rely on the face-neighbor cells. However, they are quite different in the implement details. The moment limiter approximate the high-order coefficients using the derivatives of lower order coefficients. The hierarchical reconstruction uses the mean value to do the linear reconstruction. And the successive limiting uses the secondary order reconstruction. These differences of candidates providing make the relationship between high-order and low-order coefficients quite different. The limiting of low-order coefficients has nothing to do with high-order ones in moment limiters, while the limiting of low-order coefficients relies on the high- order ones in the successive limiting. Besides, the complexity in implementation for every limiting process differs a lot. The moment limiter needs to judge the smoothness before proceeding the limiting on the same order of coefficients. It is simple but the accuracy will be degraded due to the misjudgment. The hierarchical reconstruction does not require the judgment but it needs to compute the mean value of every order before carrying out the linear reconstruction. Thus it is complicated. The successive limiting procedure uses the simple yet efficient SR to provide the candidates and requires no judgment during the limiting process. So the successive limiting seems a better choice over the other two in terms of implementation convenience and limiting effect.

In the meantime, although the stencils are large and far away from the central cell in WENO, the weights for the coefficients in the same candidate polynomials are the same. On the other hand, the successive limiting which is compact in the WBAP limiting needs to carry out the nonlinear limiting on each coefficient individually. This difference reminds us there is an intermediate way which keeps the compactness and reduces the limiting times as few as possible. And this way is going to be introduced below which is called the group weighted limiter.

For the k-th order reconstruction polynomial $u(\mathbf{x})$ on the cell Ω_i, all the k'-th order coefficients can be written as

$$u_i^l, l = l_s, \ldots, l_e, \quad l_s = DOF(k' - 1) + 1, l_e = DOF(k').$$

E.g., $k' = 3$ in the two dimensions, $l_s = 10$ and $l_e = 19$. The limiting of coefficients in this group is not taken the WBAP limiter or other TVD functions, but to use a specific weight. To suppress the numerical oscillations, this specific weight should reflect the smoothness of each candidate polynomial and only use the information in this group. According to the requirements, the weight can be defined as

$$w_i = \frac{\alpha_i}{\sum\limits_{j \in \{i, 1a, 2a, 3a\}} \alpha_j}, \alpha_j = \frac{\lambda_j}{(\beta_j + \delta)^p}, \beta_j = \sum_{l=l_s}^{l_e} (u_j^l)^2 \tag{3.40}$$

where λ_j is the linear weight. Similar to the WENO and WBAP limiters, define

$$\lambda_j = \begin{cases} \lambda_0, & j = 0 \\ 1, & \text{else.} \end{cases}$$

In the limiting process, we take $p = 2$, $\lambda_0 = 1$ as the limiting parameters. Similarly, using the successive limiting strategy, we can give the whole limiting process. Next the fourth order limiting is taken for example. For brevity, we note the candidate polynomials as

$$u_j(\mathbf{x}), j = 0, 1, 2, 3$$

where $u_o(\mathbf{x})$ represents $u_i(\mathbf{x})$, $u_j(\mathbf{x})$, $j = 1, 2, 3$ represent $u_{1a \to i}(\mathbf{x})$, $u_{2a \to i}(\mathbf{x})$ and $u_{3a \to i}(\mathbf{x})$ respectively.

(I) On the cell Ω_i, the third order coefficients for the SR can be calculated by 3.13 and they are

$$\hat{u}^l_{j\to i} = u^l_j, l = 6, 7, 8, 9, j \in \{1a, 2a, 3a\}.$$

Compute the weights on the cell $\{i, 1a, 2a, 3a\}$ using

$$w_m = \frac{\alpha_m}{\sum_{j=0}^{3}\alpha_j}, \alpha_m = \frac{\lambda_m}{(\beta_m + \delta)^p}, \beta_m = \sum_{l=6}^{9}\left(u^l_m\right)^2, m = 0, 1, 2, 3$$

Then the limited coefficients can be weighted by

$$\tilde{u}^l_i = \sum_{m=0}^{3} w_m \hat{u}^l_m, l = 6, 7, 8, 9.$$

This operation is carried out for all cells and the limited coefficients $\tilde{u}^l_i (l = 6, 7, 8, 9)$ are stored.

(II) The second order-scaled derivatives of the SRs are computed using the limited third order- scaled derivatives through

$$\begin{aligned}
\hat{u}^3_{j\to i} &= u^3_j + 3\tilde{u}^6_j \Delta x + \tilde{u}^7_j \Delta y,\\
\hat{u}^4_{j\to i} &= u^4_j + 2\tilde{u}^7_j \Delta x + 2\tilde{u}^8_j \Delta y,\\
\hat{u}^5_{j\to i} &= u^5_j + \tilde{u}^8_j \Delta x + 3\tilde{u}^9_j \Delta y,\\
\Delta x &= x_i - x_j, \Delta y = y_i - y_j, j \in \{1a, 2a, 3a\}.
\end{aligned}$$

Compute the weights on the cell $\{i, 1a, 2a, 3a\}$ using

$$w_m = \frac{\alpha_m}{\sum_{j=0}^{3}\alpha_j}, \alpha_m = \frac{\lambda_m}{(\beta_m + \delta)^p}, \beta_m = \sum_{l=3}^{5}\left(u^l_m\right)^2, m = 0, 1, 2, 3$$

Then the limited coefficients can be weighted by

$$\tilde{u}^l_i = \sum_{m=0}^{3} w_m \widehat{u}^l_m, l = 3, 4, 5.$$

This operation is carried out for all cells and the limited coefficients $\tilde{u}^l_i (l = 3, 4, 5)$ are stored.

(III) The first order-scaled derivatives of the SRs are computed using the limited second order and third order-scaled derivatives through

$$\begin{aligned}
\hat{u}^1_{j\to i} &= u^1_j + 2\tilde{u}^3_j \Delta x + \tilde{u}^4_j \Delta y + 3\tilde{u}^6_j \Delta x^2 + 2\tilde{u}^7_j \Delta x \Delta y + \tilde{u}^8_j \Delta y^2,\\
\hat{u}^2_{j\to i} &= u^2_j + \tilde{u}^4_j \Delta x + 2\tilde{u}^5_j \Delta y + \tilde{u}^7_j \Delta x^2 + 2\tilde{u}^8_j \Delta x \Delta y + 3\tilde{u}^9_j \Delta y^2,\\
\Delta x &= x_i - x_j, \Delta y = y_i - y_j, j \in \{1a, 2a, 3a\}.
\end{aligned}$$

Compute the weights on the cell $\{i, 1a, 2a, 3a\}$ using

$$w_m = \frac{\alpha_m}{\sum_{j=0}^{3} \alpha_j}, \alpha_m = \frac{\lambda_m}{(\beta_m + \delta)^p}, \beta_m = \sum_{l=1}^{2} \left(u_m^l\right)^2, m = 0, 1, 2, 3$$

Then the limited coefficients can be weighted by

$$\tilde{u}_i^l = \sum_{m=0}^{3} w_m u_m^l, l = 1, 2.$$

This operation is carried out for all cells and the limited coefficients $\tilde{u}_i^l (l = 1, 2)$ are stored.

After these three steps, all scaled derivatives are limited and the reconstruction polynomial after limiting is obtained.

3.6 Summary and Comparisons on These Three Limiters

All the three limiters introduced from Sects. 3.4–3.6 use the secondary reconstruction to provide the candidate polynomials. We analyze the reasons for constructing the three limiters and their own merits.

Since the secondary reconstruction is a complete new way for providing the candidate polynomials, we do not know much about its effect thus the mature and widely used WENO limiter is used as the limiting function and limiting process. The construction of k-exact WENO limiter is to validate the efficiency of secondary reconstruction, including the stencils choosing and accuracy analysis. Since the computation of WENO smooth indicator will be complicated on unstructured grids, the quadrature-free implementation is used. Thus, we have the easy and low cost k-exact WENO limiter.

Even the optimized k-exact WENO limiter has its own defects, e.g., the candidate stencils is large and far away from the central cell and the smooth indicator is complex in high-order schemes, and these defects would become more obvious in mixed element and higher dimensions. Thus, the construction of WBAP limiters is to deal with the limiting stencils and limiting function problems, which are solved through the proposal of WBAP limiting function and successive limiting procedure. And from the numerical cases in Sect. 3.9, we will see the WBAP limiters can achieve similar resolutions and shock capturing abilities as WENO limiters and take less computational time. In the meantime, the WBAP limiter provides a new framework of constructing the unstructured limiters, in which different biased functions and weighted function can achieve different limiting effects. So the WBAP limiter is significant to the development of accuracy preserving limiters on unstructured grids. We will continue to study its behaviors further.

The successive weighted limiter is aimed to keep the compactness of WBAP limiter and reduce the limiting times. It can be viewed as a modification of WBAP limiter. From the limiting effect, the WBAP limiter can suppress numerical oscillations near the strong shock even using conservative variables, while the successive weighted limiter must rely on the expensive transformation to characteristic variables. From the development of limiters, the WBAP limiter is more like the TVD limiter thus it can be applied on other limiters using TVD functions. And successive weighted limiter can be viewed as the compact way to realize the WENO limiter. It is interesting to find that for the second order limiting, the successive weighted limiter is identical to the WENO limiter. This means the successive weighted limiter is a generalization of second order WENO limiter to high-order schemes in an efficient way.

These three limiters can be further optimized in the current framework, e.g., the linear weights in k-exact WENO limiters, the parameters in WBAP function and the weights in the successive weighted limiter. Among these possible optimizations, we point out that the dissipation for the high-order coefficients can be less than that of lower order ones. And choosing different weights for coefficients of different orders can give more accurate results while capturing the numerical oscillations.

3.7 The Problem-Independent Shock Detector

The purpose of the proposed limiters is to suppress the oscillations near the discontinuities. In the regions away from the discontinuities, the limiting procedure actually is unnecessary. Moreover, the limiters are always computationally expensive especially with the characteristic transformation even when we propose the more efficient three limiters. E.g., the fourth order k-exact WENO limiters take nearly two-third of the total time if the limiters are applied on every cell. Thus a shock detector is always adopted to distinguish the smooth cells from the cells near discontinuities, i.e., the trouble cells. This technique significantly improves the computational efficiency and is extensively applied on the high-order numerical schemes, e.g., the DG and SV methods [9, 23]. Since the accuracy preserving limiters are hard to design in the DG and SV schemes, the shock detectors are crucial to the accuracy of the computational results, which will be degraded once the smooth cells are mistaken for the trouble cells. Presently, all kinds of shock detectors cannot only mark all the cells near shock waves as trouble cells. The shock detector described below is only used for reducing the computational time since the three shock capturing schemes can be very accurate even in the smooth regions. The overall accuracy of the schemes will not be affected.

The present shock detector relies on the smoothness indicator IS_i defined by

$$IS_i = \frac{\sum_j \left| u_i(\mathbf{x}_i) - u_j(\mathbf{x}_i) \right|}{N_j h^{(k+1)/2} \max(\bar{u}_j, \bar{u}_i)}, \quad j \in \{1a, 2a, 3a\} \tag{3.41}$$

where $\mathbf{x}_i$ is the centroid of cell Ω_i, $u_i(\mathbf{x})$ is the PR polynomial and $u_j(\mathbf{x})$, $j = 1a, 2a, 3a$ are the PRs on the direct face neighbors of Ω_i, as depicted in Fig. 2.2. N_j stands for the number of first neighborhood cells, for triangular cell $N_j = 3$. In terms of the secondary reconstructions, we can further modify Eq. (3.41) into

$$IS_i = \frac{\sum_j \left|u_i(\mathbf{x}_i) - u_{j\to i}(\mathbf{x}_i)\right|}{N_j h^{(k+1)/2} \max(\bar{u}_j, \bar{u}_i)}, \quad j \in \{1a, 2a, 3a\} \tag{3.42}$$

which is computationally more efficient because of the specific forms of the primary and secondary reconstructions. However, this specific form of smoothness indicator is only applicable for third and higher order schemes because we have $u_i(\mathbf{x}_i) = u_{j\to i}(\mathbf{x}_i) = \bar{u}_i$ for the second order scheme.

Since only the point value at the centroid of the central cell is needed to evaluate at the numerator in Eq. (3.42), this shock detector can be a bit more efficient than the KXRCF shock detector [24] for the DG method, which needs the integration over the interface of the central cell. And the denominator in Eq. (3.42) is almost the same as KXRCF detector except we take into account for the number of first neighborhood cells. So analogy to the analyze of KXRCF detector, it is easy to distinguish the smooth region from the region near discontinuities as follows

$$u_i(\mathbf{x}_i) - u_{j\to i}(\mathbf{x}_i) = \begin{cases} O(h^{k+1}), & \text{in smooth region} \\ O(h^1), & \text{near discontinuity} \end{cases}.$$

Therefore, as the cell size $h \to 0$, $IS_i = 0$ in smooth regions and $IS_i = \infty$ near the discontinuities. So we define

$$(IS_i < \bar{S}_{\text{dis}}) = \begin{cases} \text{true}, & \text{smooth region} \\ \text{false}, & \text{shock region} \end{cases}$$

We generally choose $\bar{S}_{\text{dis}} = 1$ for the second and third order schemes, and $\bar{S}_{\text{dis}} = 3$ for the fourth order scheme. These parameters work well for all test cases of the present paper and are problem independent.

3.8 Numerical Accuracy Test

The superiority of high-order limiters over low-order ones is that, the high-order accuracy still remains even the limiter is applied on all cells. This section is going to test the accuracy of the k-exact WENO limiters, WBAP limiters and group weighted limiters through different numerical examples. The parameters in the limiters are chosen conservative ones which are beneficial for shock capturing. Even those conservative parameters can give results as accurate as the unlimited ones. The CFL number is taken to be 0.8 for all cases and no shock detector is used in the computation.

3.8.1 Isentropic Vortex Problem

The isentropic vortex transport problem [7, 16] is used to examine the accuracy of the numerical scheme in computing multidimensional flow without shock waves. The mean flow is $\rho = 1$, $p = 1$,and $(u, v) = (1, 1)$. We add, to mean flow, an isentropic vortex expressed by the following perturbations,

$$(\delta u, \delta v) = \frac{\varepsilon}{2\pi} e^{0.5(1-r^2)} (-\bar{y}, \bar{x})$$

$$\delta T = -\frac{(\gamma - 1)\varepsilon^2}{8\gamma\pi^2} e^{1-r^2}, \quad \delta S = 0$$

with $(\overline{x}, \overline{y}) = (x - 5, y - 5)$, $r^2 = \overline{x}^2 + \overline{y}^2$, and the vortex strength $\varepsilon = 5$.

We compute the solution at $t = 2.0$ to test the accuracy of the proposed schemes without the shock detector. Two types of grids, namely regular grids and irregular grids which are shown as shown in Fig. 3.8, are used for test the capabilities of the two types of limiters on different grids. The second to fourth order FV schemes are used in the computations. $\lambda_0 = 100$ is used for k-exact WENO limiters on the emphasis of central cell, while $\lambda_0 = 10$ for WBAP limiters and $\lambda_0 = 1$ for group weighted limiter. In the next section, we will see these parameters are sufficient to suppress the numerical oscillations of strong shock waves.

The numerical results are shown in Table 3.2 for the unlimited FV scheme on the two types of grids, Table 3.3 for the k-exact WENO schemes, Table 3.4 for the WBAP limiters and Table 3.5 for the group weighted limiters. In terms of convergence rate of the density, all schemes can achieve their theoretical values in terms of L_1 norm and L_∞ norm. Moreover, even if the limiters are applied on all cells, the schemes still remain high-order accuracy and the errors of limited schemes are smaller than that of unlimited ones sometimes.

3.8.2 The Constructed Exact Solution of Three Dimensional Euler Equations

The description of the constructed problem is in Sect. 2.5.1. This section will test the accuracy of the k-exact WENO and WBAP limiters by solving the three- dimensional Euler equations. The parameter to emphasize the central cell is chosen as $\lambda_0 = 100$ and $\lambda_0 = 10$ and the Riemann solver is Roe. The computational results are shown in Table 3.6, where we can observe that all limiters can achieve the expected orders. The L_1 errors of WBAP limiters are smaller than that of k-exact WENO schemes, but L_∞ error of WBAP limiter are larger.

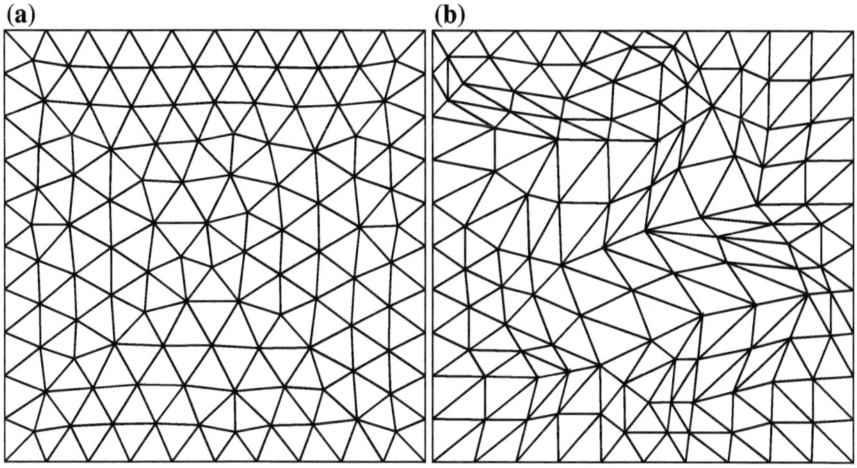

Fig. 3.8 Regular and Irregular grids used for accuracy test of isentropic vortex. Reprinted from Ref. [22], Copyright 2012, with permission from Elsevier

Table 3.2 Accuracy test for isentropic vortex using unlimited FV scheme

		Regular grids				Irregular grids			
Schemes	Grid	L_1 error	Order	L_∞ error	Order	L_1error	Order	L_∞ error	Order
2nd order	1/10	6.16E-03	–	7.94E-02	–	1.50E-02	–	0.237029	–
	1/20	1.57E-03	1.98	2.23E-02	1.83	4.97E-03	1.59	9.70E-02	1.29
	1/40	3.52E-04	2.15	5.91E-03	1.91	1.26E-03	1.98	1.98E-02	2.29
	1/80	8.19E-05	2.10s	1.34E-03	2.15	2.60E-04	2.28	4.06E-03	2.28
	1/160	2.07E-05	1.99	3.16E-04	2.08	5.73E-05	2.18	1.09E-03	1.9
3rd order	1/10	5.71E-03	–	9.65E-02	–	1.14E-02	–	0.209066	–
	1/20	1.01E-03	2.49	2.55E-02	1.92	3.36E-03	1.75	8.51E-02	1.3
	1/40	1.61E-04	2.65	3.47E-03	2.88	7.31E-04	2.2	2.14E-02	1.99
	1/80	1.89E-05	3.09	3.89E-04	3.16	1.18E-04	2.63	3.79E-03	2.5
	1/160	2.40E-06	2.98	5.09E-05	2.93	1.58E-05	2.91	5.16E-04	2.88
4th order	1/10	3.48E-03	–	5.32E-02	–	9.96E-03	–	0.156396	–
	1/20	3.88E-04	3.16	7.83E-03	2.76	2.05E-03	2.28	5.09E-02	1.62
	1/40	2.42E-05	4.00	4.85E-04	4.01	2.86E-04	2.84	8.18E-03	2.64
	1/80	1.22E-06	4.31	2.36E-05	4.36	1.93E-05	3.89	6.53E-04	3.65
	1/160	7.46E-08	4.03	1.67E-06	3.82	9.56E-07	4.33	3.29E-05	4.31

3.9 Numerical Tests

In last section, the accuracy tests are carried out for the three limiters and all of them can achieve the expected ones. This section is continued to test the shock capturing ability, resolution, smoothness, and robustness for these limiters. Without specified, the CFL number is taken as 0.8.

Table 3.3 Accuracy test for isentropic vortex using the k-exact WENO limiters

		Regular grids				Irregular grids			
Schemes	Grid	L_1 error	Order	L_∞ error	Order	L_1 error	Order	L_∞ error	Order
2nd order	1/10	7.37E-03	–	0.154133	–	1.28E-02	–	0.242832	
	1/20	1.81E-03	2.03	4.34E-02	1.83	4.78E-03	1.42	0.12587	0.95
	1/40	4.34E-04	2.06	8.13E-03	2.42	1.43E-03	1.74	3.71E-02	1.76
	1/80	9.13E-05	2.25	1.70E-03	2.26	3.61E-04	1.99	1.21E-02	1.62
	1/160	2.17E-05	2.07	4.71E-04	1.85	7.96E-05	2.18	2.65E-03	2.18
3rd order	1/10	6.95E-03	–	0.118129	–	1.33E-02	–	0.242519	–
	1/20	9.71E-04	2.84	2.45E-02	2.27	3.93E-03	1.76	9.79E-02	1.31
	1/40	1.65E-04	2.56	3.64E-03	2.75	8.81E-04	2.16	2.41E-02	2.03
	1/80	1.95E-05	3.08	4.03E-04	3.17	1.46E-04	2.6	4.50E-03	2.42
	1/160	2.43E-06	3.01	5.31E-05	2.92	1.94E-05	2.91	6.18E-04	2.86
4th order	1/10	3.64E-03	–	6.44E-02	–	9.83E-03	–	0.17502	–
	1/20	3.40E-04	3.42	6.83E-03	3.24	2.09E-03	2.23	4.83E-02	1.86
	1/40	2.20E-05	3.95	5.12E-04	3.74	2.64E-04	2.99	7.29E-03	2.73
	1/80	1.11E-06	4.31	2.06E-05	4.63	1.58E-05	4.06	5.10E-04	3.84
	1/160	7.00E-08	3.99	1.50E-06	3.78	7.90E-07	4.32	2.52E-05	4.34

Table 3.4 Accuracy test for isentropic vortex using WBAP limiters

		Regular grids				Irregular grids			
Schemes	Grid	L_1 error	Order	L_∞ error	Order	L_1 error	Order	L_∞ error	Order
2nd order	1/10	8.79E-03	–	0.216589	–	1.41E-02	–	0.300499	
	1/20	2.73E-03	1.69	5.74E-02	1.92	6.22E-03	1.18	0.155547	0.95
	1/40	5.98E-04	2.19	1.98E-02	1.53	1.75E-03	1.83	5.18E-02	1.59
	1/80	1.19E-04	2.33	4.44E-03	2.16	4.87E-04	1.85	1.99E-02	1.38
	1/160	2.39E-05	2.31	9.59E-04	2.21	1.12E-04	2.12	6.15E-03	1.69
3rd order	1/10	1.05E-02	–	0.232484	–	1.59E-02	–	0.300639	–
	1/20	2.08E-03	2.34	5.54E-02	2.07	6.35E-03	1.33	0.14183	1.08
	1/40	1.53E-04	3.76	4.32E-03	3.68	1.03E-03	2.63	2.88E-02	2.3
	1/80	1.29E-05	3.57	4.24E-04	3.35	8.72E-05	3.56	3.85E-03	2.91
	1/160	1.38E-06	3.22	4.81E-05	3.14	7.33E-06	3.57	3.55E-04	3.44
4th order	1/10	8.08E-03	–	1.78E-01	–	1.42E-02	–	0.276498	–
	1/20	4.26E-04	4.25	1.07E-02	4.05	2.96E-03	2.26	5.63E-02	2.3
	1/40	2.14E-05	4.32	3.98E-04	4.75	1.77E-04	4.06	5.30E-03	3.41
	1/80	9.91E-07	4.43	2.06E-05	4.27	9.13E-06	4.28	3.48E-04	3.93
	1/160	5.87E-08	4.08	1.48E-06	3.8	5.63E-07	4.02	2.66E-05	3.71

3.9.1 Rotation of Slotted Cylinder

This case is the two-dimensional benchmark problem proposed by LeVeque [25] which makes it possible to asses the ability of a high-resolution scheme to pre-

Table 3.5 Accuracy test for isentropic vortex using group weighted limiters

		Regular grids				Irregular grids			
Schemes	Grid	L_1 error	Order	L_∞ error	Order	L_1 error	Order	L_∞ error	Order
3rd order	1/10	6.94E-03	–	0.149426	–	1.35E-02	–	0.265009	–
	1/20	9.59E-04	2.86	2.30E-02	2.7	3.99E-03	1.76	9.60E-02	1.46
	1/40	1.41E-04	2.76	3.20E-03	2.84	7.23E-04	2.46	2.03E-02	2.24
	1/80	1.58E-05	3.16	3.39E-04	3.24	1.02E-04	2.82	3.22E-03	2.65
	1/160	1.97E-06	3	4.48E-05	2.92	1.25E-05	3.03	4.17E-04	2.95
4th order	1/10	4.45E-03	–	9.43E-02	–	1.05E-02	–	2.12E-01	–
	1/20	2.58E-04	4.11	5.51E-03	4.1	1.93E-03	2.44	4.86E-02	2.12
	1/40	1.62E-05	3.99	3.69E-04	3.9	2.03E-04	3.25	5.40E-03	3.17
	1/80	8.52E-07	4.25	1.88E-05	4.3	1.06E-05	4.25	3.28E-04	4.04
	1/160	5.22E-08	4.03	1.15E-06	4.02	4.82E-07	4.46	1.55E-05	4.4

Table 3.6 Accuracy test for the three-dimensional Euler equations using k-exact WENO and WBAP limiters

	Schemes	Grid	L_1 error	Order	L_∞ error	Order
k-exact WENO	3rd order	1/10	3.02E-03		6.27E-03	
		1/20	3.75E-04	3.01	7.79E-04	3.01
		1/30	1.11E-04	3.01	2.28E-04	3.03
		1/40	4.67E-05	3	9.65E-05	2.99
	4th order	1/10	2.29E-04	–	6.68E-04	–
		1/15	4.28E-05	4.13	1.30E-04	4.03
		1/20	1.34E-05	4.04	4.06E-05	4.05
		1/25	5.44E-06	4.03	1.66E-05	4
WBAP limiter	3rd order	1/10	2.96E-03		7.70E-03	
		1/20	3.63E-04	3.03	9.86E-04	2.97
		1/30	1.08E-04	3	2.94E-04	2.99
		1/40	4.54E-05	3	1.24E-04	2.99
	4th order	1/10	2.06E-04		8.19E-04	
		1/15	3.98E-05	4.05	1.76E-04	3.79
		1/20	1.29E-05	3.91	4.96E-05	4.41
		1/25	5.38E-06	3.93	2.24E-05	3.56

serve both smooth and discontinuous profile. The problem is governed by the scalar advection equation, i.e.,

$$u_t + \mathbf{v} \cdot \nabla u = 0,$$

with the nonuniform velocity field $\mathbf{v} = (0.5 - y, x - 0.5)$. This problem describes that a slotted cylinder, a sharp cone and a smooth hump undergo a counterclockwise rotation about the center of a unit square $[0, 1] \times [0, 1]$. The shapes of the three bodies are depicted in Fig. 3.9 and further details of the initial conditions can be found in [25].

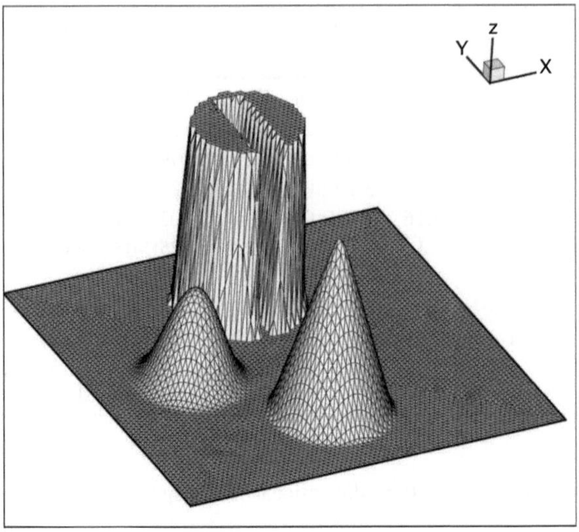

Fig. 3.9 Slotted cylinder rotation: initial data/exact solution

For one period rotation at $t = 2\pi$, the exact solution is the initial condition. Figure 3.10 shows the computational results on the grids of size $h = 1/160$ using different orders of the three limiters. The parameter on emphasizing the central cell are chosen to be $\lambda_0 = 100$, $\lambda_0 = 10$ and $\lambda_0 = 1$ for k-exact WENO limiters, WBAP limiters and group weighted limiters, respectively. To compare the resolutions for different limiters on the smooth regions, the velocity distributions at $x = 0.5$ and $y = 0.75$ are drawn and shown in Fig. 3.11. We can see no oscillations occur using these three limiters. And higher order schemes have better solutions in capturing the discontinuities and come closer to exact solutions on smooth regions.

3.9.2 Shock Tube Problems

This test case is chosen to test the capability of the three limiters in capturing the strong shock wave and contact discontinuity. The problem is one-dimensional in nature, and the initial conditions are

$$(\rho, u, p)(\mathbf{x}) = \begin{cases} (\rho_L, u_L, p_L) & \text{if } x \leq 0.5 \\ (\rho_R, u_R, p_R) & \text{if } x > 0.5 \end{cases}$$

We consider two sets of initial conditions given by

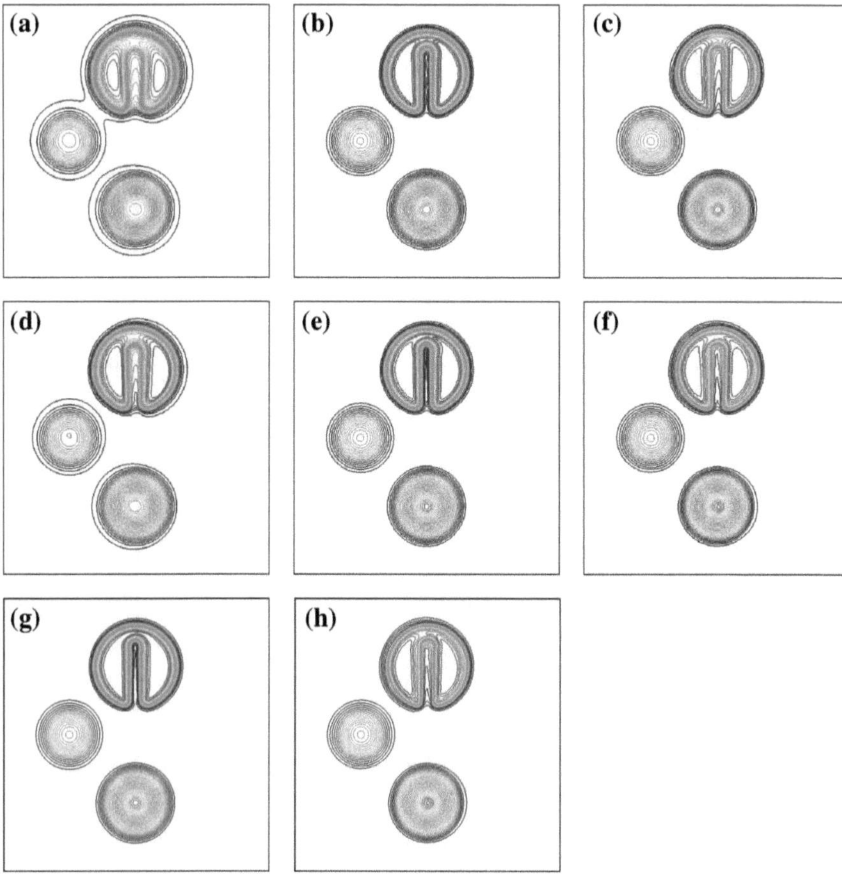

Fig. 3.10 Contour comparison of the slotted cylinder at $t = 2\pi$

Case 1:

$$\begin{aligned}(\rho_L, u_L, p_L) &= (1., 0.75, 1.),\\ (\rho_R, u_R, p_R) &= (0.125, 0., 0.1)\\ t_{\text{end}} &= 0.2\end{aligned}$$

Case 2:

$$\begin{aligned}(\rho_L, u_L, p_L) &= (0.445, 0.698, 3.528),\\ (\rho_R, u_R, p_R) &= (0.5, 0, 0.571)\\ t_{\text{end}} &= 0.1\end{aligned}$$

where Case 1 is the Sod tube problem [26] and Case 2 is the Harten-Lax problem [27]. Although the problems are one-dimensional, they are computed using both two-dimensional grids on region $[0, 1] \times [0, 0.2]$ with grid size $h = 1/200$, which are shown in Fig. 3.12. To test the sensitivity of the parameter on the emphasis of central

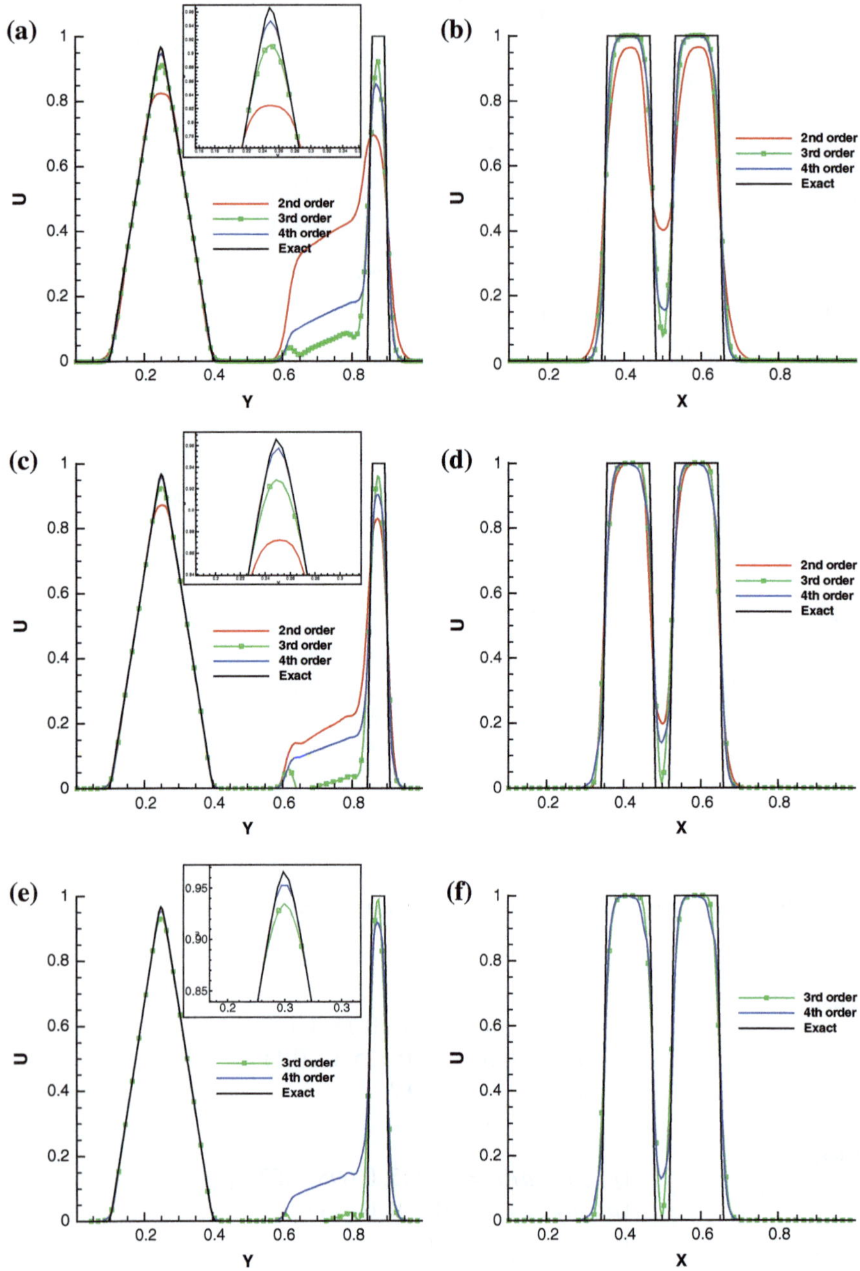

Fig. 3.11 The solutions along *vertical* and *horizontal lines* for the slotted cylinder at $t = 2\pi$

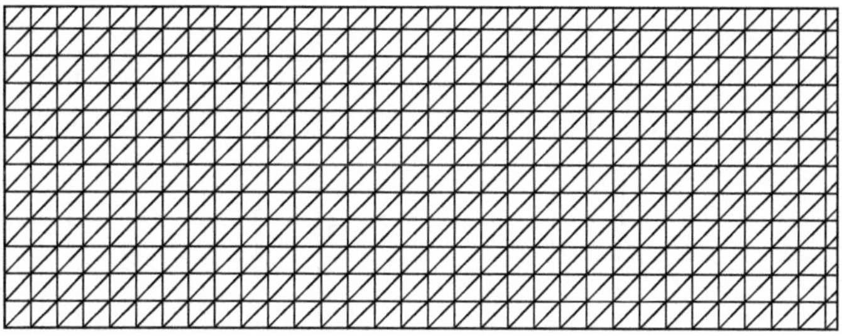

Fig. 3.12 Two-dimensional grids for shock tube problem

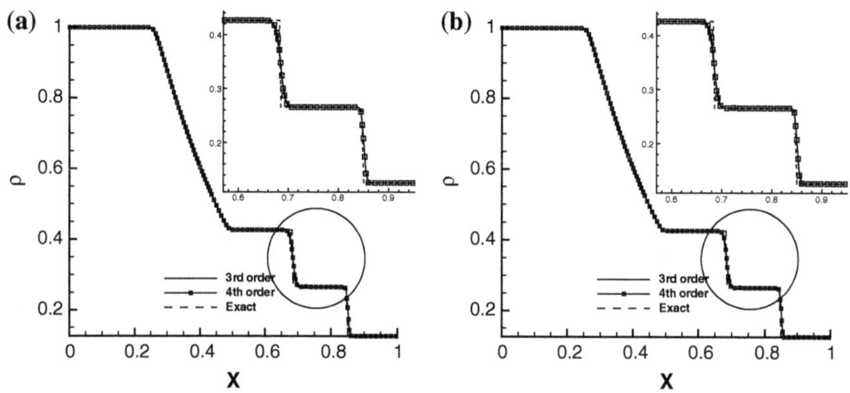

Fig. 3.13 Comparison of density distributions of Sod tube using k-exact WENO limiters

cell, two sets of parameter are chosen, namely $\lambda_0 = 100$ and 1000 for k-exact WENO limiter, $\lambda_0 = 1$ and 10 for WBAP limiter and $\lambda_0 = 1$ for group weighted limiter. Roe scheme is used as the Riemann solver. The computational results are shown in Figs. 3.13, 3.14, 3.15, 3.16, and 3.17, where we can observe that all schemes can suppress the numerical oscillations near the shock waves and contact discontinuities with different parameters.

3.9.3 Shu-Osher Problem

This test case is to solve the Shu and Osher [28] problem which describes the interaction of an entropy sine wave with a Mach 3 righting moving shock. The domain is taken as $[0, 10] \times [0, 1]$ and the initial conditions are

$$(\rho, u, p) = \begin{cases} (3.857143, 2.629369, 10.33333), & x < 1, \\ (1 + 0.2\sin(5x), 0, 1) & x \geq 1. \end{cases}$$

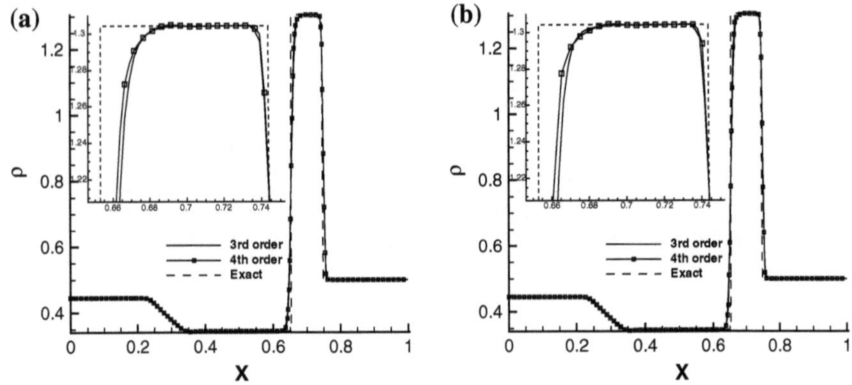

Fig. 3.14 Comparison of density distributions of Lax tube using k-exact WENO limiters

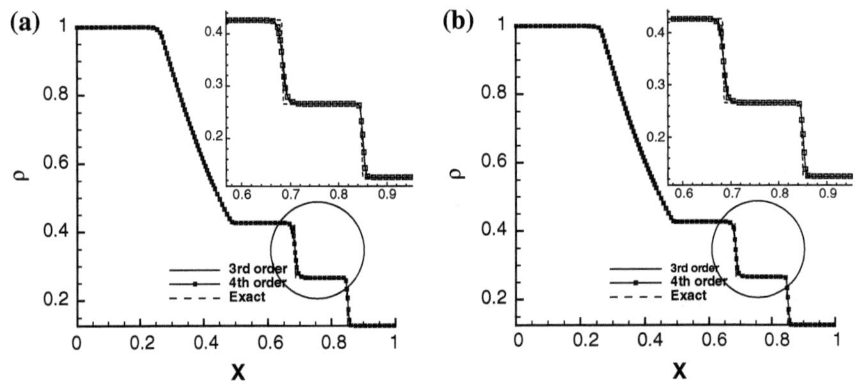

Fig. 3.15 Comparison of density distributions of Sod tube using WBAP limiters

The grids used for computation is uniform triangular grids as shown in Fig. 3.12 with grid size $h = 1/40$, namely 400 grids along the x direction. The simulation is run to $t = 1.8$. Since the exact solution for this problem is unknown, we consider the one-dimensional numerical solution computed by the fifth order WENO scheme [29] using 20,000 cells as the exact solution. This case is used to test the resolutions for the three limiters. The second to fourth order FV schemes are used with the HLL Riemann solver. No shock detector is used in the computation.

The results using the three limiters are shown in Figs. 3.18, 3.19, and 3.20, where the left is the global view and the right is the enlarged view for the interaction of shock wave and sine wave. The three limiters are capable to capture the shock waves in the flow field. And with the raise of order, the results come closer to exact solution. The fourth order scheme even matches the extrema of the exact solution. For the three limiters in the scheme of the same order, their results are close to each other.

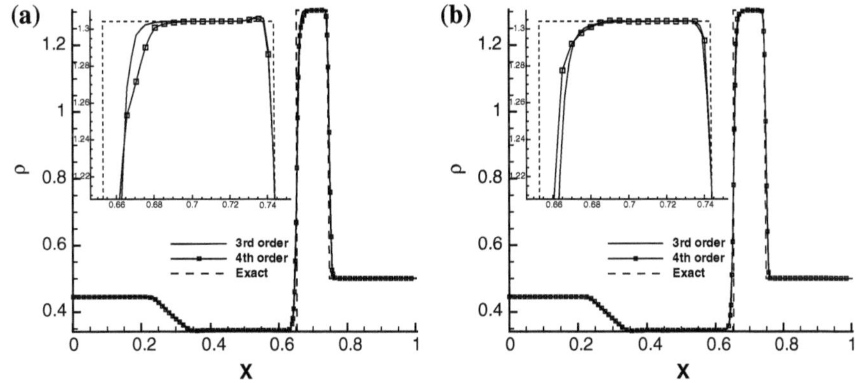

Fig. 3.16 Comparison of density distributions of Lax tube using WBAP limiters

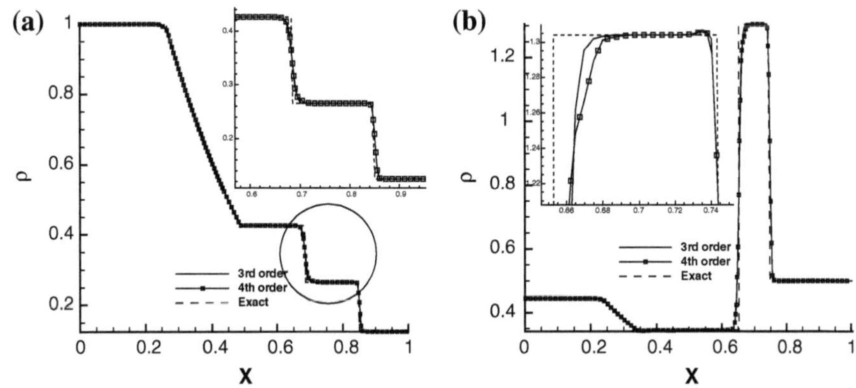

Fig. 3.17 Comparison of density distributions of Sod and Lax tube using group weighted limiters

3.9.4 Double Mach Reflection Problem

One of the popular test cases for high-resolution schemes is the double Mach reflection problem [30]. The whole computational domain is $[0, 4] \times [0, 1]$. The wall is located at the bottom of the computational domain beginning at $x = 1/6$. Initially, a right-moving shock with $Ma = 10$ is located at $x = 1/6$, $y = 0$, inclined $60°$ with respect to the x-axis. The computational is carried out until $t = 0.21$.

The FV schemes with different limiters are used to capture the shock waves and the density contours are shown in Figs. 3.21 and 3.22. We can observe that the three limiters are capable to capture the shock waves and no obvious numerical oscillations generate in the flow field. Higher order schemes can capture more complicated structures in the flow field.

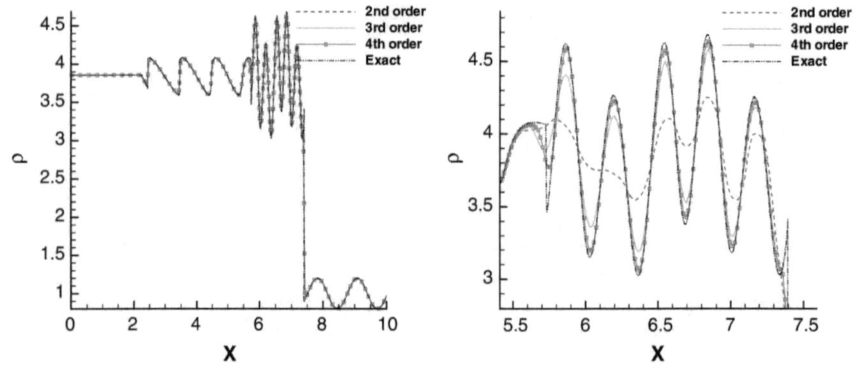

Fig. 3.18 Comparison of density distributions of Shu-Osher problem using k-exact WENO limiters

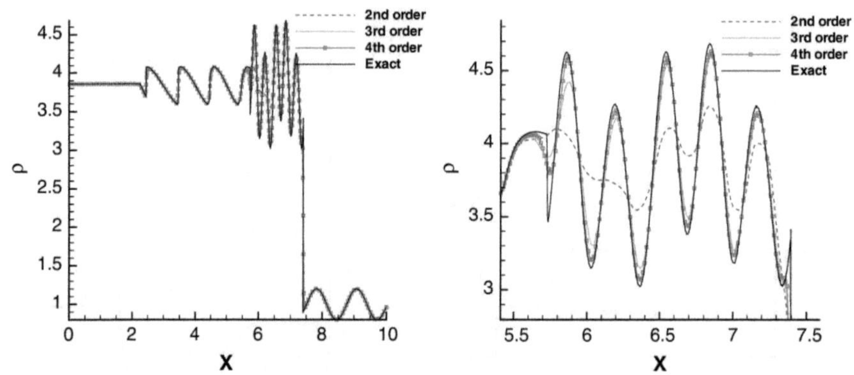

Fig. 3.19 Comparison of density distributions of Shu-Osher problem using WBAP limiters

3.9.5 A Mach 3 Wind Tunnel with a Step

This is another popular case for high resolution schemes used by Woodward and Colella [30]. The computational domain is $[0, 3] \times [0, 1]$. The corner of the step is located at $(x, y) = (0.6, 0.2)$. The initial conditions are $(\rho, u, v, p) = (1.4, 3, 0, 1)$ which stand for a Mach 3 uniform flow impacting the step at the initial time.

To avoid over expansion, the grids near the corner of wind tunnel are refined as shown in Fig. 3.23. Since the grids at the corner is refined and the global time step is small, the problem-independent shock detector is used in the simulation to further improve the efficiency of the simulation. Two set of grids, $h = 1/80$ and $h = 1/160$, are used in the computation and the results are shown in Figs. 3.24, 3.25, and 3.26. All these three limiters can capture the shock waves without oscillations. Higher order schemes show higher resolutions in capturing the shock waves. The fourth order even predicts the roll up of the contact discontinuities.

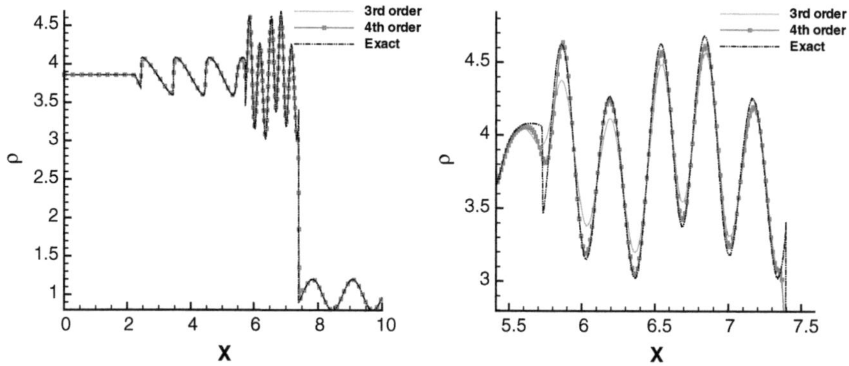

Fig. 3.20 Comparison of density distributions of Shu-Osher problem using group weighted limiters

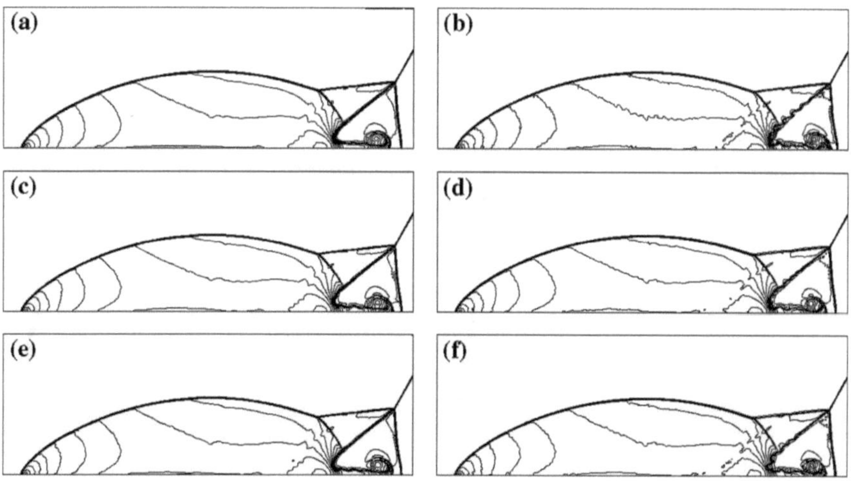

Fig. 3.21 Comparison of density contours for double Mach reflection on girds with size $h = 1/240$. Thirty equally spaced *contour lines* from ρ =2.0 to 21.7

3.9.6 Transonic Flow Around NACA0012 Airfoil

Transonic steady flow about the NACA0012 airfoil is simulated to demonstrate the superior convergence property of the WBAP limiters for steady flow computation. A comparison is also made with the k-exact WENO scheme to test the efficiency of both schemes. The unstructured grid used in the simulation is shown in Fig. 3.27, which consists of 10,382 elements, 5,306 grid points, and 150 boundary points.

The far field condition of test case is the Mach 0.8 transonic inflow at the angle of 12.5°. We compute this test case with the third and fourth order WBAP and k-exact WENO schemes. The WBAP-L2 limiter is used with $n = 10$. Since straight segments approximation for curved boundary would degrade the accuracy and stall the

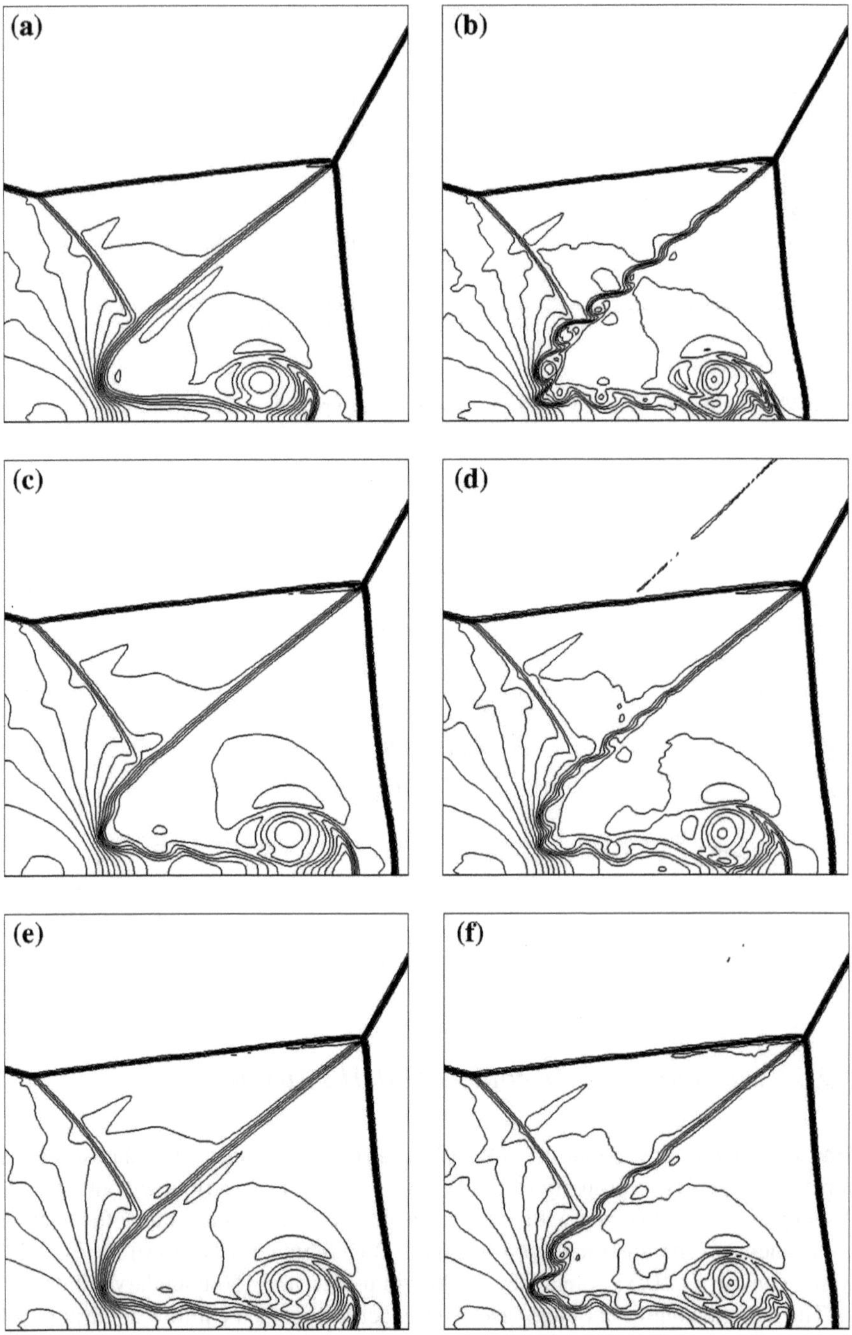

Fig. 3.22 Close-up view around the double Mach stem of Fig. 3.21

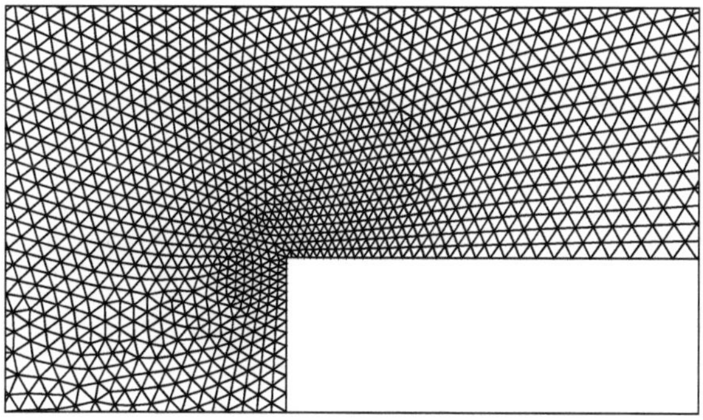

Fig. 3.23 Unstructured mesh used for computing supersonic flow in a wind tunnel with a step at $M_\infty = 3$. Reprinted from Ref. [22], Copyright 2012, with permission from Elsevier

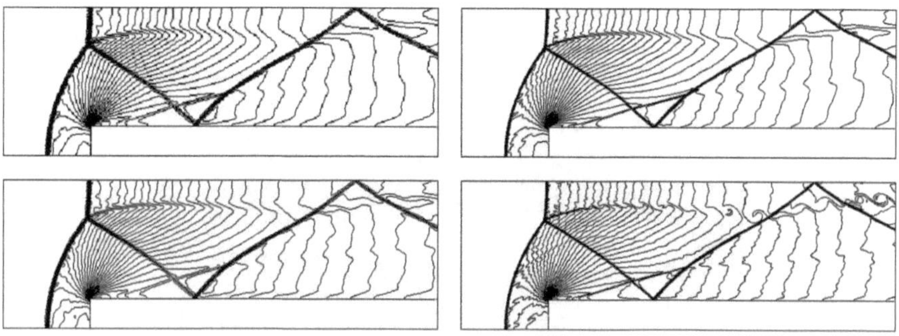

Fig. 3.24 Comparison of density contours for the Mach 3 wind tunnel with a step using k-exact WENO limiters. *Left*, $h = 1/80$; *right*, $h = 1/160$. *Up*, third order; *down*, fourth order

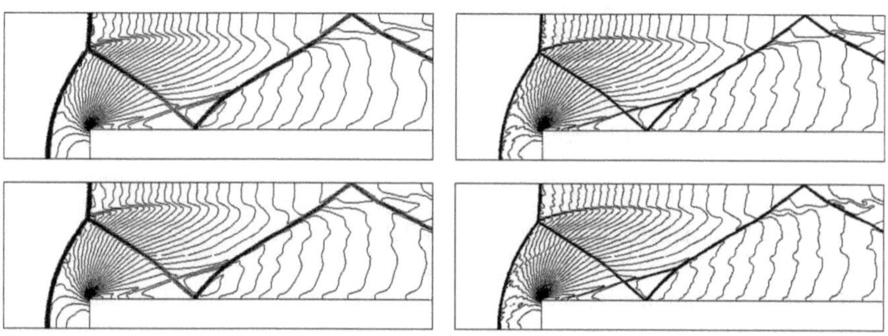

Fig. 3.25 Comparison of density contours for the Mach 3 wind tunnel with a step using WBAP limiters. *Left*, $h = 1/80$; *right*, $h = 1/160$. *Up*, third order; *down*, fourth order

convergence of the high-order methods, the method in Chap. 4 is used to treat the curved boundary. A comparison of the convergence histories is shown in Fig. 3.28. Both the third order WBAP and k-exact WENO schemes can converge to machine zero, while the residuals can drop about four orders for the fourth order WBAP

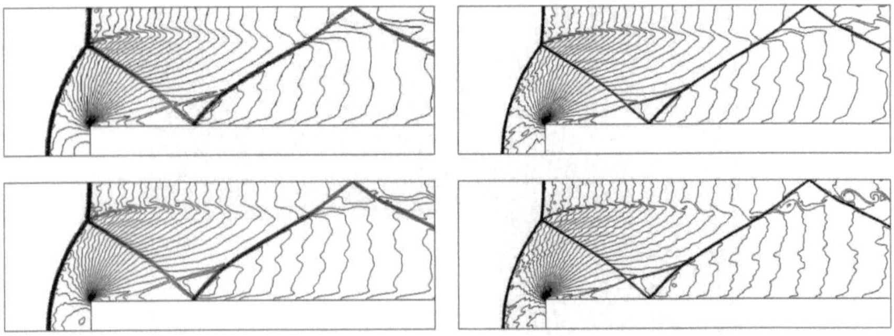

Fig. 3.26 Comparison of density contours for the Mach 3 wind tunnel with a step using group weighted limiters. *Left*, $h = 1/80$; *right*, $h = 1/160$. *Up*, 3rd order; *down*, 4th order

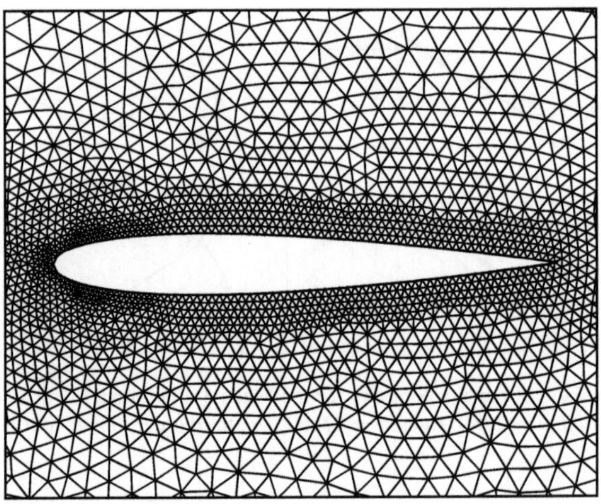

Fig. 3.27 Unstructured mesh ($n_{elem} = 10,382$, $n_{point} = 5,306$, $n_{bound} = 150$) used for computing transonic and supersonic flow past a NACA0012 airfoil. Reprinted from Ref. [22], Copyright 2012, with permission from Elsevier

scheme and 3.5 orders for the fourth order WENO scheme. From the pressure distributions shown in Fig. 3.29, it is observed that both the third order WBAP and k-exact WENO schemes can resolve the shock waves without oscillations. The fourth order WBAP scheme yields a slightly sharper shock transition but generates a small overshoot near the upper shock wave. In other regions, the solutions of the WBAP and k-exact WENO schemes are very close.

To compare the computational efficiency of both WBAP and k-exact WENO schemes, we audit the CPU time used in 200 iterations which is shown in Table 3.7. It can be seen clearly that the WBAP schemes are considerably more efficient than the k-exact WENO schemes.

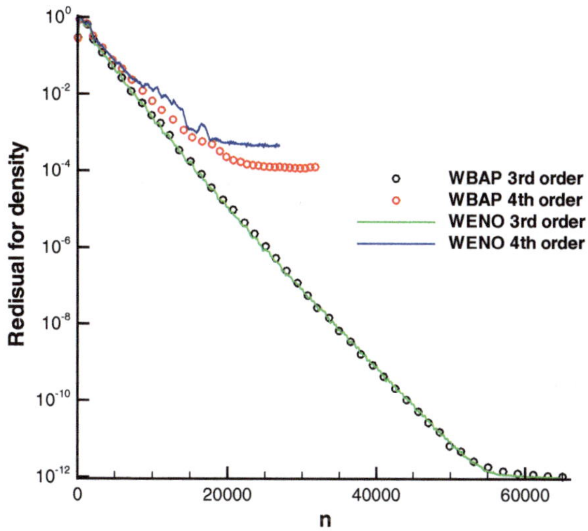

Fig. 3.28 The convergence history of transonic flow around NACA0012 airfoil using the third and fourth order WBAP and k-exact WENO schemes. Reprinted from Ref. [22], Copyright 2012, with permission from Elsevier

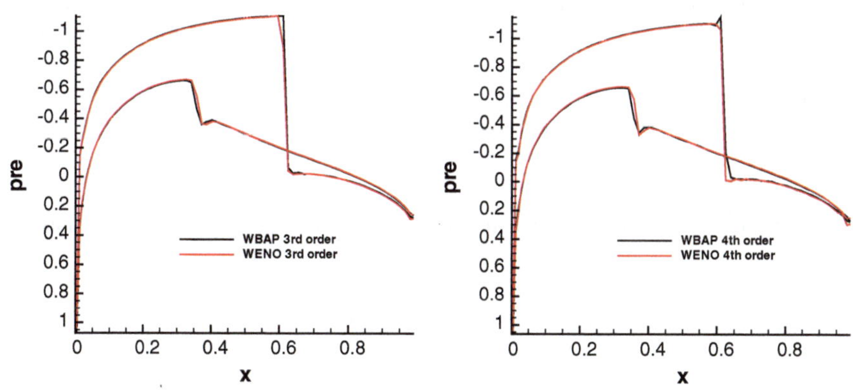

Fig. 3.29 Distributions of surface pressure coefficient using the WBAP and k-exact WENO schemes. Reprinted from Ref. [22], Copyright 2012, with permission from Elsevier

Table 3.7 Computational time comparison between WBAP and k-exact WENO schemes for 200 iterations. Reprinted from Ref. [22], Copyright 2012, with permission from Elsevier

	WBAP		WENO			
Time (s)	third order	fourth order	third order, QF	third order	fourth order, QF	fourth order
Total	8.8	16.3	12.8	20.3	29.2	52.8
Limiter	3.1	7.3	7.1	14.6	20.2	43.8

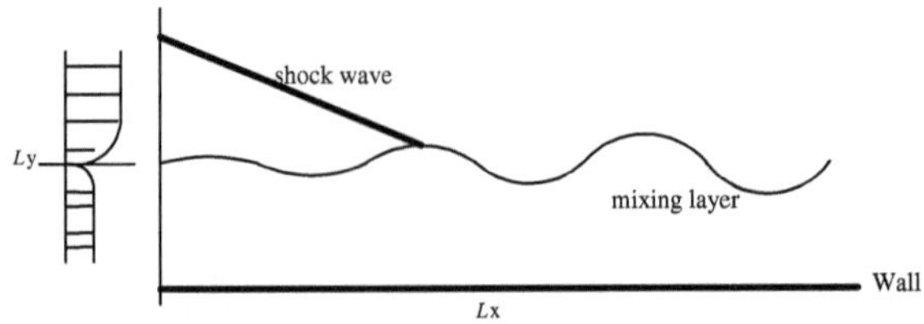

Fig. 3.30 The schematic diagram for the spatially evolving mixing layer

3.9.7 Shock Wave Impingement on a Spatially Evolving Mixing Layer

The cases above are inviscid flows and in this section we reproduce the example presented in [31], which is the viscous flows with shock waves. This case is developed to test the behavior of the schemes for shock waves interacting with shear layers where the vortices arising from the shear layer instability are forced to pass through a shock wave, as shown in Fig. 3.30. Although the resulting shock wave is not strong, shock-capturing strategy is still necessary to suppress the oscillations near the shock. The problem domain is the rectangle $0 \le x \le 200$ and $-20 \le y \le 20$ with 401 uniform grids along the x direction and 101 stretched grids in y direction which can be referred to [31].

The inflow is specified with a hyperbolic tangent profile,

$$u = 2.5 + 0.5\tanh(2y)$$

and a periodic fluctuation to vertical component of the velocity inflow, given as

$$v' = \sum_{k=1}^{2} a_k \cos(2\pi kt/T + \phi_k)\exp(-y^2/b)$$

where $b = 10$ and $T = \frac{\lambda}{u_c}$, wavelength $\lambda = 30$, convective velocity $u_c = 2.68$. For $k = 1$ we take $a_1 = 0.05$ and $\phi = 0$, and for $k = 2$ we take $a_2 = 0.05$ and $\phi = \pi/2$. No perturbations are added to the u-component of velocity. For the flow is supersonic at the outflow, no specific boundary condition is needed. On the lower boundary, a slip wall condition is specified. With the problem setup, an oblique shock originates from the top left corner, impacting the shear layer around $x = 90$. The shock wave reflects at the lower wall and passes back through the deflected shear layer. The Reynolds number is chosen to be 500.

The problem is computed using second to fourth order k-exact WENO schemes. The results are shown at $t = 120$. The density contours are depicted in Fig. 3.31, where we can observe that the vortex structure generated by the mixing layer is more clear with higher order schemes. The pressure and temperature contours using fourth

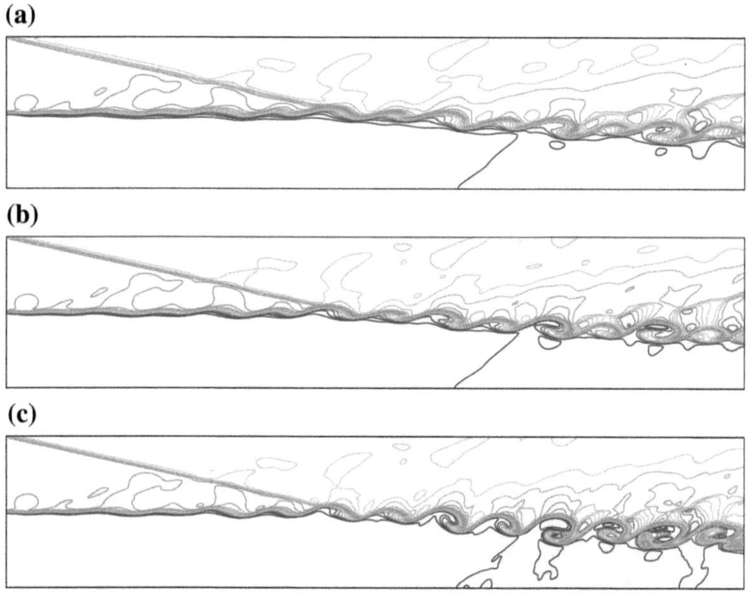

Fig. 3.31 The density *contour lines* of the interaction between shock waves and mixing layer at $Tt120$ using different schemes

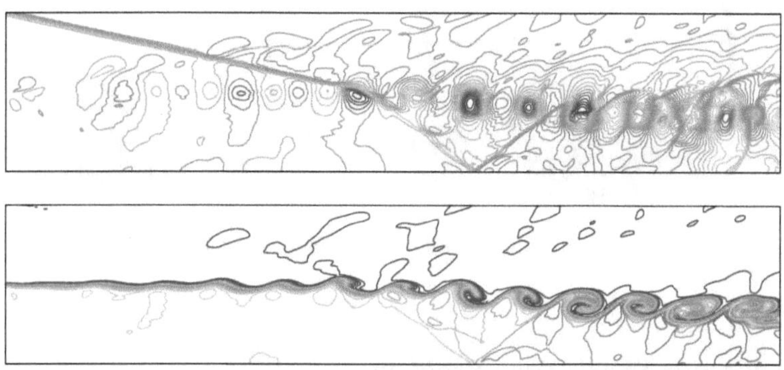

Fig. 3.32 The pressure (*up*) and temperature (*down*) *contour lines* of the interaction between shock waves and mixing layer at $t = 120$

order scheme are also shown in Fig. 3.32. The weak shock waves and the rolling-up of vortex are finely captured using the high order scheme.

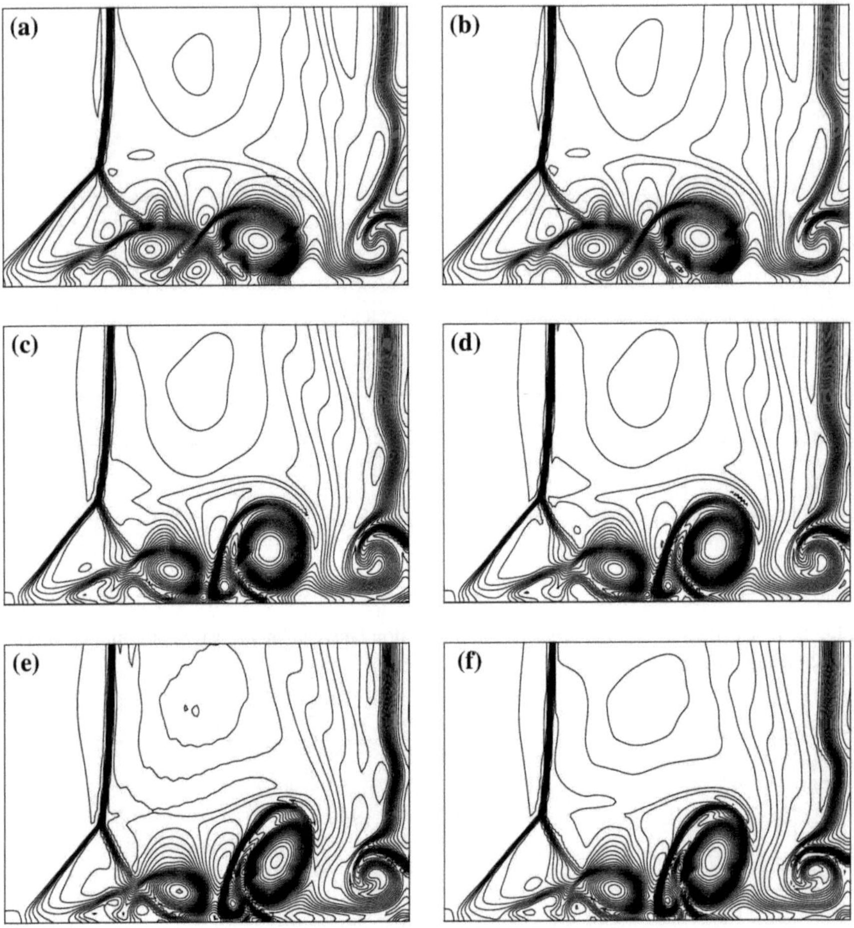

Fig. 3.33 The *contour lines* of viscous shock tube problems using different schemes on grids with size $h = 1/250$

3.9.8 Viscous Shock Tube Problem

This test case shows viscous flow structure accompanying the interaction between shock wave, boundary layer, and vortex. λ-shape shock wave and vortices can be observed by the interaction between the viscous boundary layer at the horizontal wall and the reflective shock wave from the vertical wall. Computational domain is 1 length unit wide and 1 length unit high, and the diaphragm is located at $x = 0.5$. The initial condition is given as follows

$$(\rho_L, u_L, v_L, p_L) = (120, 0, 0, 120/\lambda), (\rho_R, u_R, v_R, p_R) = (1.2, 0, 0, 1.2/\lambda).$$

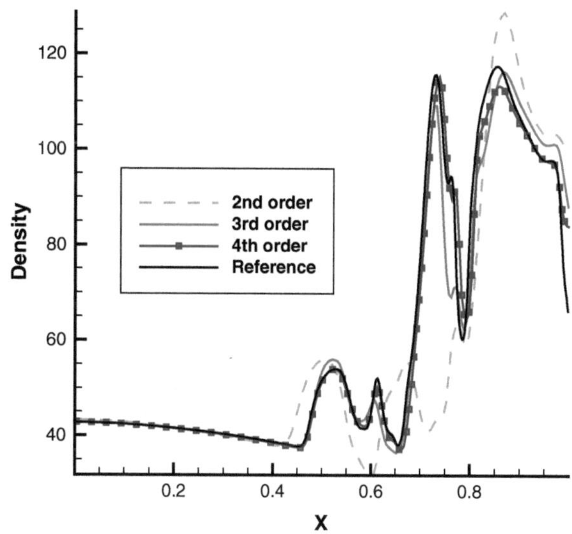

Fig. 3.34 The comparison of density at the *bottom* of viscous shock tube using WBAP limiters of different orders

The Reynolds number is 200 and the Prandtl number is 0.73.

This problem is crucial to validate the strong shock capturing ability in the viscous flow field. The mesh we use is uniform triangular grids with grid size $h = 1/250$. In Fig. 3.33 density contours computed by second to fourth order WBAP and k-exact WENO limiters are compared at $t = 1$. The higher order limiters can get more clear λ-shape vortices. Figure 3.34 compares the density distribution along the bottom wall using WBAP limiters, where a third order WENO limiter on finer grids $h = 1/500$ is used for reference. The result of higher order WBAP limiter comes closer to the reference values in most parts.

3.9.9 Explosion Problems on Two and Three Dimensions

This section will compute the two- and three-dimensional explosion problem to test the shock capturing ability of the limiters. The governing equations is the Euler equations and the region is the circle of sphere of radius $R = 1$. The initial conditions are

$$(\rho, \mathbf{v}, p)(\mathbf{x}, 0) = \begin{cases} (1., \mathbf{0}, 1.0), & \text{if } \| x \| < 0.4, \\ (0.125, \mathbf{0}, 0.1), & \text{else.} \end{cases}$$

This test case can be viewed as the extension of one-dimensional Sod tube problem. The reference results can be computed using the one-dimensional Euler equations, i.e.,

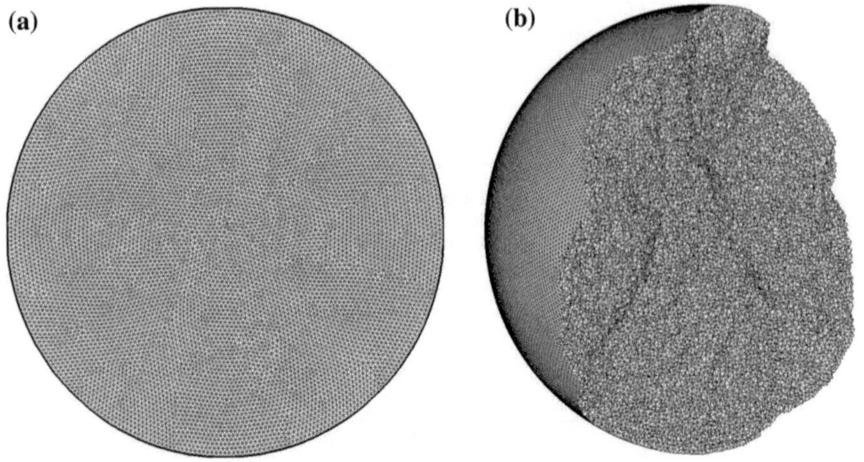

Fig. 3.35 Two- and three-dimensional grids for explosion problem

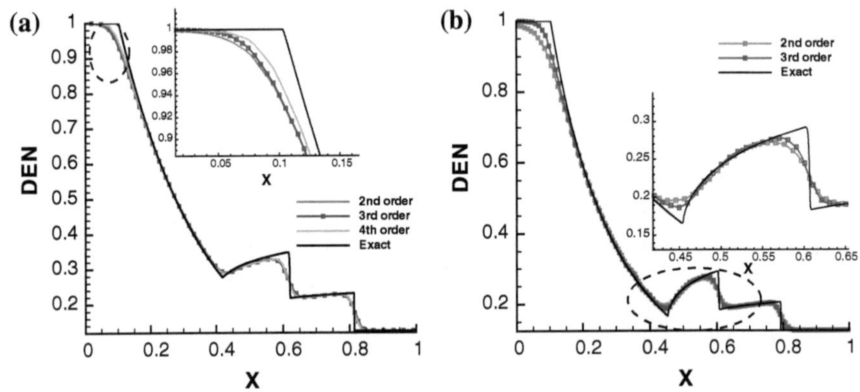

Fig. 3.36 Density distribution on the positive x-axis for the 2D and 3D explosion problem using k-exact WENO scheme

$$U_t + F(U)_r = S(U)$$

with the source term $S(U) = -\alpha/r \cdot \left[\rho u, \rho u^2, u(E+p)\right]$, $\alpha = 1$ and 2 for two and three dimensions, respectively. The computational grids are the triangles or tetrahedrons with grid size $h = 0.02$, as shown in Fig. 3.35. We solve the problem using the second to fourth order k-exact WENO schemes in two dimensions and the second to third order schemes in three dimensions. From Fig. 3.36, we can see that in both two and three dimensions no oscillations occur near the discontinuities and rarefaction and high-order schemes can obtain better accuracy than lower order one.

3.10 Conclusions

The chapter continues the development of high-order FVM and summarizes a set of high-order limiters construction method, which consists of the candidate polynomial construction, limiting function, and limiting process. For all these three parts, the chapter proposes some novel high-efficient solution strategies. First, the SR is proposed to construct the candidate polynomials. As a continuation of k-exact reconstruction, the SR is much more convenient for implement and more efficient comparing to the traditional multiple reconstructions. To validate this efficiency, the mature WENO limiter is used to provide the limiting function and limiting procedure. In the meanwhile, the quadrature-free way is used to compute the WENO smooth indicator and the limiting stencils are optimized. Thus, we have the so-called k-exact WENO limiter. Second, to make the limiting compact, a novel WBAP limiter is created. This limiter improved the BAP function to WBAP function, which is smoother and adjustable in dissipations. The WBAP function is simplified and its nonoscillatory suppressing mechanism can be theoretically analyzed. With the successive limiting procedure, the WBAP limiter is used to limiting the coefficients of the polynomials from the highest orders and the lowest ones. This WBAP limiter is high- order accuracy preserving limiter which is compact and easy implemented and gives a new way to construct the high-order limiters. Third, to reduce the limiting times of WBAP limiter, we construct the group weighted limiter. Abundance of numerical examples show that the three limiters proposed in this chapter not only suppress the numerical oscillations near the discontinuities, but also keep the high- order accuracy and high resolution in the smooth regions. For the complicated shock waves in the viscous flows, the nonoscillatory FVM also gives satisfied results.

References

1. Godunov SK (1959) A finite-difference method for the numerical computation of discontinuous solutions of the equations of fluid dynamics. Mat Sbornik 47:271–306
2. Harten A, Engquist B, Osher S et al (1987) Uniformly high order accurate essentially non-oscillatory schemes (III). J Comput Phys 71:231–303
3. Harten A (1991) Recent development in shock-capturing schemes. ICASE report No. 91–8
4. Zhang XX, Shu CW (2010) On maximum-principle-satisfying high order schemes for scalar conservation laws. J Comput Phys 229:3091–3120
5. Barth TJ, Jespersen D (1989) The design and application of upwind schemes on unstructured meshes. In: Proceedings of the 27th AIAA aerospace sciences meeting, Reno, NV, Paper AIAA 89–0366
6. Venkatakrishnan V (1995) Convergence to steady state solutions of the Euler equations on unstructured grids with limiters. J Comput Phys 118:120–130

7. Park JS, Yoon SH, Kim C (2010) Multi-dimensional limiting process for hyperbolic conservation laws on unstructured grids. J Comput Phys 229:788–812
8. Park JS, Kim C (2011) Higher-order discontinuous Galerkin-MLP methods on triangular and tetrahedral grids. In: AIAA 2011–3059, 20th AIAA computational fluid dynamics conference, Honolulu, Hawaii, 27–30 June 2011
9. Cockburn B, Shu CW (2001) Runge-Kutta discontinuous Galerkin methods for convection-dominated problems. J Sci Comput 16:173–261
10. Luo H, Baum JD, Löhner R (2007) A hermite WENO-based limiter for discontinuous Galerkin method on unstructured grids. J Comput Phys 225:686–713
11. Yang M, Wang ZJ (2009) A parameter-free generalized moment limiter for high-order methods on unstructured grids. Adv Appl Math Mech 1(4):451–480
12. Li WA, Ren YX. High order k-exact WENO finite volume schemes for solving gas dynamic Euler equations on unstructured grids. Int J Numer Meth Fluids DOI:10.1002/fld.2710
13. Friedrich O (1998) Weighted essentially non-oscillatory schemes for the interpolation of mean values on unstructured grids. J Comput Phys 144:194–212
14. Dumbser M, Käser M (2007) Arbitrary high order non-oscillatory finite volume schemes on unstructured meshes for linear hyperbolic systems. J Comput Phys 221:693–723
15. Dumbser M, Käser M, Titarev VA et al (2007) Quadrature-free non-oscillatory finite volume schemes on unstructured meshes for nonlinear hyperbolic systems. J Comput Phys 226:204–243
16. Hu C, Shu CW (1999) Weighted essentially non-oscillatory schemes on triangular meshes. J Comput Phys 150:97–127
17. Sweby PK (1984) High-resolution schemes using flux limiters for hyperbolic conservation laws. SIAM J Numer Anal 21(5):995–1011
18. Choi H, Liu JG (1998) The reconstruction of upwind fluxes for conservation laws: its behavior in dynamic and steady state calculations. J Comput Phys 144:237–256
19. Zoppou C, Roberts S (2003) Explicit schemes for dam-break simulations. J Hydraul Eng 129:11–34
20. Cada M, Torrilhon M (2009) Compact third-order limiter functions for finite volume methods. J. Comput. Phys. 228:4118–4145
21. Li WA, Ren YX, Lei GD et al (2011) The multi-dimensional limiters for solving hyperbolic conservation laws on unstructured grids. J Comput Phys 230(21):7775–7795
22. Li WA, Ren YX (2012) The multi-dimensional limiters for solving hyperbolic conservation laws on unstructured grids II: extension to high order finite volume schemes. J Comput Phys 231(11):4053–4077
23. Wang ZJ (2002) Spectral (finite) volume method for conservation laws on unstructured grids, basic formulation. J Comput Phys 178:210–251
24. Krivodonova L et al (2004) Shock detection and limiting with discontinuous Galerkin methods for hyperbolic conservation laws. Appl Numer Math 48:323–338
25. LeVeque RJ (1996) High-resolution conservative algorithms for advection in incompressible flow. Siam J Numer Anal 33:627–665
26. Sod G (1978) A survey of several finite difference methods for systems of nonlinear hyperbolic conservation laws. J Comput Phys 27:1–31
27. Lax PD (1954) Weak solutions of nonlinear hyperbolic equations and their numerical computation. Commun Pure Appl Math 7:159–193
28. Shu CW, Osher S (1988) Efficient implementation of essentially non-oscillatory shock capturing schemes. J Comput Phys 77:439–471

29. Jiang GS, Shu CW (1996) Efficient implementation of weighted ENO schemes. J Comput Phys 126:202–228
30. Woodward P, Colella P (1984) The numerical simulation of two-dimensional fluid flow with strong shocks. J Comput Phys 54:115–173
31. Yee HC, Sandham ND, Djomehri MJ (1999) Low-dissipative high-order shock-capturing methods using characteristic-based filters. J Comput Phys 150:199–238

Chapter 4
Mixed Element and Curved Boundary Treatment

In the above computations, we use the two-dimensional triangular and three-dimensional tetrahedron for the shape of control volumes. Generally speaking, the triangle and tetrahedron are more suitable for generating quasi-isotropic grids. However, the flow change is quite different in every direction for different types of flows, e.g., to solve the boundary layer in viscous flows, quadrilateral grids rather than triangular grids are more efficient to capture the large gradients in the wall normal direction. For the second order FVM, there is no special requirement for the shape of control volume. The polygon or polyhedron can be used in second order FVM. However, since the k-exact reconstruction and Gaussian quadrature are needed in the high-order FVM, the implement is quite different on different types of grids. And the high-order quadrature rules are extremely difficult on polygon and polyhedron. Thus, only the simplex are considered in this chapter, i.e., two-dimensional triangle and quadrilateral, three-dimensional tetrahedron, pyramid, prism, and hexahedron grids which are shown in Figs. 4.1 and 4.2. The key is the face and volume integrals on these types of grids. The method to carry out the integrals is to transform these nonregular elements to regular ones using the isoparametric transformation.

Besides the integrals on the simplex, the isoparametric transformation can also be used to treat the curved boundary problem which is considered as another challenge in high-order schemes. For the second order FVM, all boundaries can be represented using straight-side elements, which agree with the linear representation inside the elements. However, in high-order schemes, the straight-side element cannot give enough information to represent the high-order accuracy for the curved elements. So, the main work is to represent the curved surface with parameters, e.g., more control points on the curved surface. Then the isoparametric transformation is used to transform the curved cells to regular cells for the face and volume integrals. Nevertheless, different isoparametric transformations require different computational cost. This chapter is to make the isoparametric transformation as simple as possible and carries on the integrals more efficiently.

Through the isoparametric transformation, the face and volume integrals can be carried out in regular regions. Using the methods on the former two chapters, we can

W. Li, *Efficient Implementation of High-Order Accurate Numerical Methods on Unstructured Grids*, Springer Theses, DOI: 10.1007/978-3-662-43432-1_4,

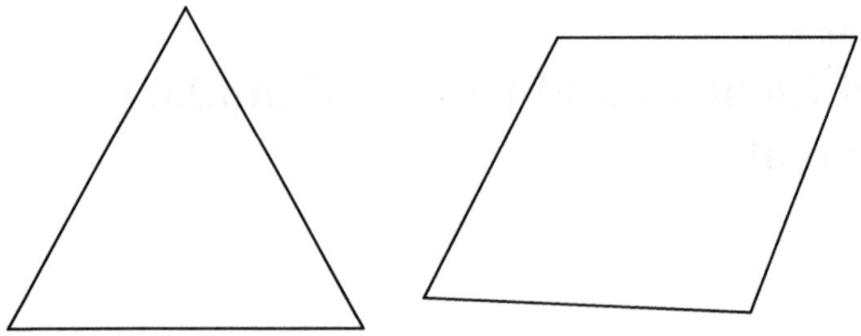

Fig. 4.1 Triangular and quadrilateral elements

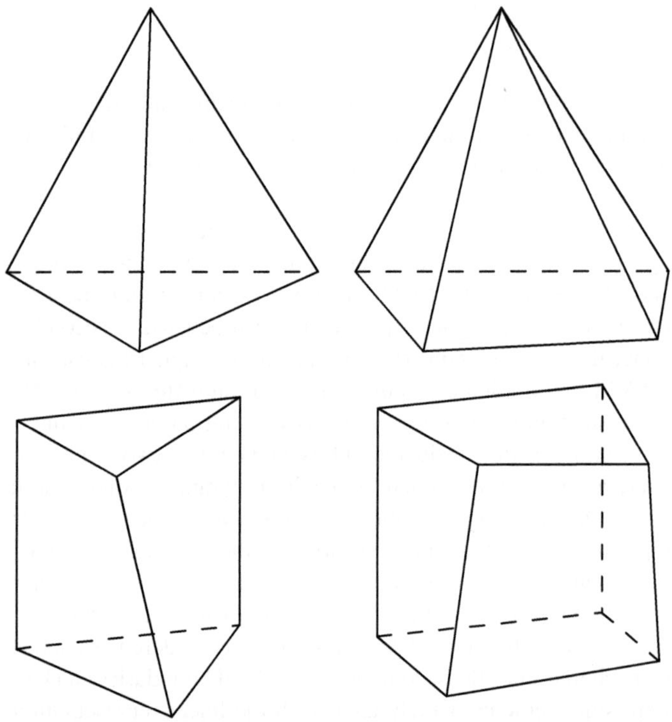

Fig. 4.2 Tetrahedral, pyramid, prism, and tetrahedral elements from *left* to *right*, *up* to *bottom*

generalize the high-order FVM to mixed elements and curved boundaries by solving the stencils choosing problems for the k-exact reconstruction and the implement of limiters. Finally, the high-order FVM is validated using a series of complicated flows.

4.1 The Applications of Isoparametric Transformation on Mixed Elements and Curved Boundaries

To implement the FVM on mixed elements and curved boundaries, we first introduce the isoparametric transformation on mixed elements and curved boundaries. Using the Lagrangian interpolation, the isoparametric transformation can be expressed as

$$\mathbf{x} = \sum_{k=1}^{N} \mathbf{x}_k v_k\left(\xi, \eta, \zeta\right) \tag{4.1}$$

where (ξ, η, ζ) is the coordinate in the isoparametric element, N is the number of control points on physical coordinate with $\mathbf{x}_k$ being the coordinates, and $v_k(\xi, \eta, \zeta)$ is the basis function which satisfies the following relation

$$v_i\left(\xi_j\right) = \begin{cases} 1, & i = j \\ 0, & i \neq j \end{cases}$$

with the parametric coordinate ξ_j corresponding to the physical coordinate x_j. Besides, to make the isoparametric transformation coincide on the cell interface for the adjacent cells, the isoparametric transformation still needs to satisfy the following condition,

$$\text{if } \left(\xi, \eta, \zeta\right) \in S, \quad \text{then } \forall \mathbf{x}_k \notin S, v_k\left(\xi, \eta, \zeta\right) = 0 \tag{4.2}$$

This condition means the isoparametric transformation on the face/edge only relies on the local control points on the face/edge. When we use the simplified isoparametric transformation, this condition can be used to verify the correctness of the transformation. The following sections will choose suitable control points and basis functions for the simplified isoparametric transformation on the mixed grids and curved surfaces.

4.1.1 The Isoparametric Transformation on Mixed Elements

In order to carry on the integration in high-order FVM, the control volumes shown in Figs. 4.1 and 4.2 must be transformed to regular regions using the isoparametric transformation. The one on two-dimensional quadrilateral grids can be shown in Fig. 4.3, which can be written as the following bilinear form,

$$\begin{aligned} v_1 &= (1-\xi)(1-\eta), \\ v_2 &= \xi(1-\eta), \\ v_3 &= \xi\eta, \\ v_4 &= \eta(1-\xi), \end{aligned} \tag{4.3}$$

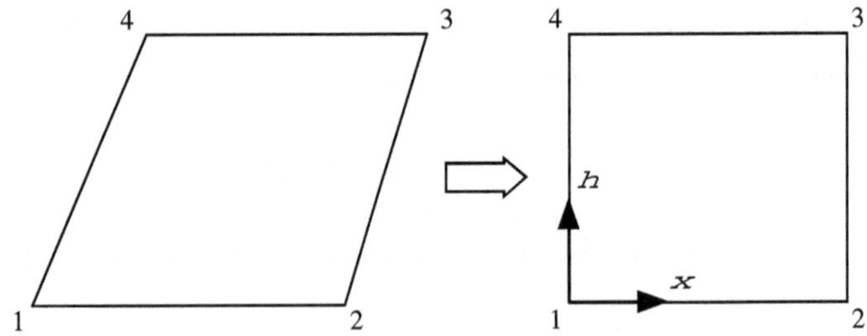

Fig. 4.3 The isoparametric transformation on the quadrilateral grid

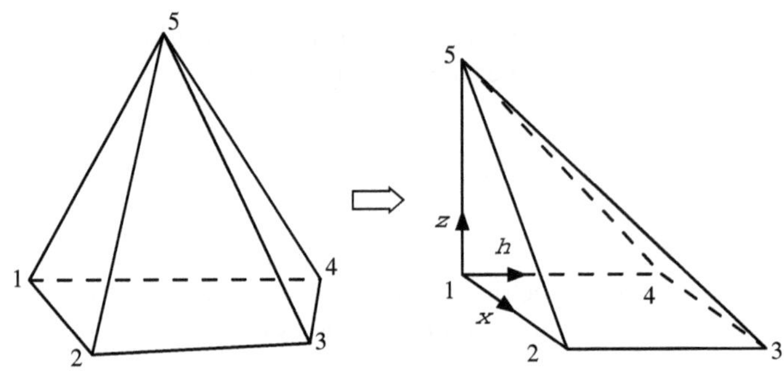

Fig. 4.4 The isoparametric transformation on the pyramid grid

The quadrilateral is changed to regular regions $[0, 1] \times [0, 1]$ and the two-dimensional Gaussian quadrature can be chosen as the tensor product of one-dimensional quadrature.

For the three-dimensional grids, i.e., the pyramid, prism, and hexahedron, all need to be transformed to the isoparametric element. The strategy shown in Fig. 4.4 for pyramid elements can be written as

$$\begin{aligned}
v_1 &= (1 - \xi - \zeta)(1 - \eta - \zeta), \\
v_2 &= \xi(1 - \eta - \zeta), \\
v_3 &= \xi\eta, \\
v_4 &= \eta(1 - \xi - \zeta), \\
v_5 &= \zeta.
\end{aligned}$$

The best way to integrate over the pyramid grid is dividing the grid into two tetrahedron grids. The strategy shown in Fig. 4.5 for prism elements can be written as

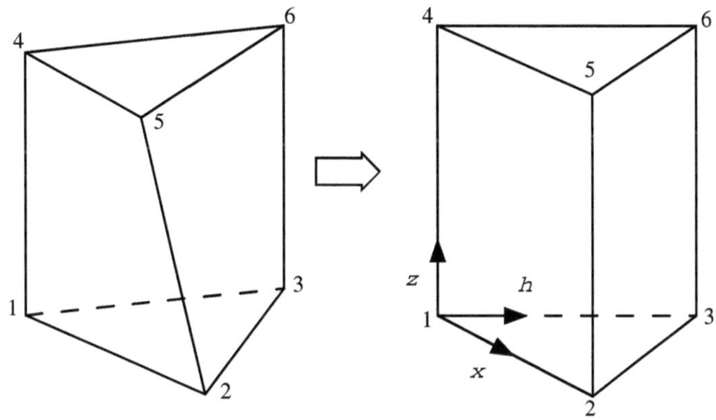

Fig. 4.5 The isoparametric transformation on the prism grid

$$
\begin{aligned}
v_1 &= (1-\xi-\eta)(1-\zeta),\\
v_2 &= \xi(1-\zeta),\\
v_3 &= \eta(1-\zeta),\\
v_4 &= \zeta(1-\xi-\eta),\\
v_5 &= \xi\zeta,\\
v_6 &= \eta\zeta.
\end{aligned}
$$

And the integration over the prism grid is the tensor product of one-dimensional Gaussian quadrature in the ζ direction and two-dimensional Gaussian quadrature on $\xi-\eta$ plane. The strategy shown in Fig. 4.6 for hexahedron elements can be written as

$$
\begin{aligned}
v_1 &= (1-\xi)(1-\eta)(1-\zeta),\\
v_2 &= \xi(1-\eta)(1-\zeta),\\
v_3 &= \xi\eta(1-\zeta),\\
v_4 &= \eta(1-\xi)(1-\zeta),\\
v_5 &= \zeta(1-\xi)(1-\eta),\\
v_6 &= \xi\zeta(1-\eta),\\
v_7 &= \xi\eta\zeta,\\
v_8 &= \eta\zeta(1-\xi).
\end{aligned}
$$

And the integration over the hexahedron grid is the tensor product of one-dimensional Gaussian quadrature in ξ, η, ζ directions, respectively.

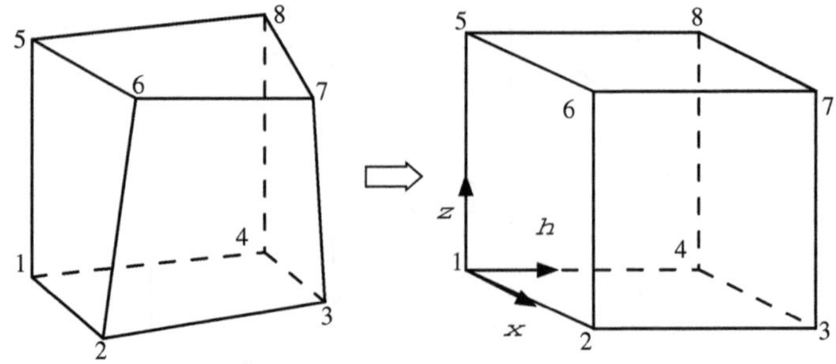

Fig. 4.6 The isoparametric transformation on the hexahedron grid

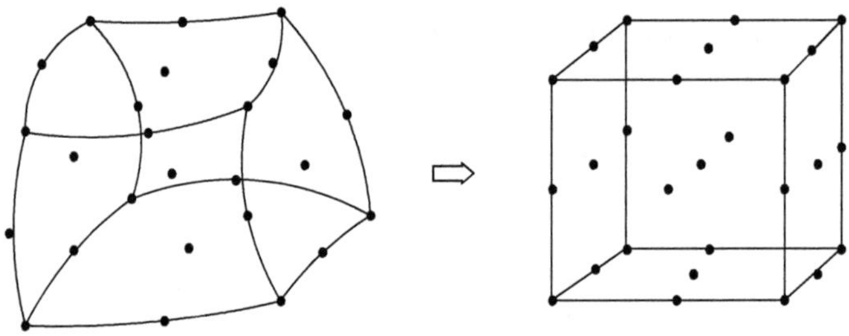

Fig. 4.7 The transformation for the *curved* hexahedron grid

4.1.2 The Isoparametric Transformation on Curved Elements

The traditional transformation for the curved tetrahedron and hexahedron grids requires lots of control points as shown in Fig. 4.7. The complete transformation is very complicated. For most cases, each control volume only contains one curved boundary. We only know the coordinates of the control points on the surface and the inner points inside the cells are unknown. This means we need to simplify the traditional transformation to one using only the control points on the curved surface and vertex. This chapter only considers elements with one curved surface. Those with more than one curved surfaces will be divided into serval smaller cells with one curved surface.

For the two-dimensional grids, we only give the isoparametric transformation for the curved triangular and quadrilateral grids using the third order polynomial approximation as shown in Figs. 4.8 and 4.9. The isoparametric transformation for triangle shown in Fig. 4.8 can be written as

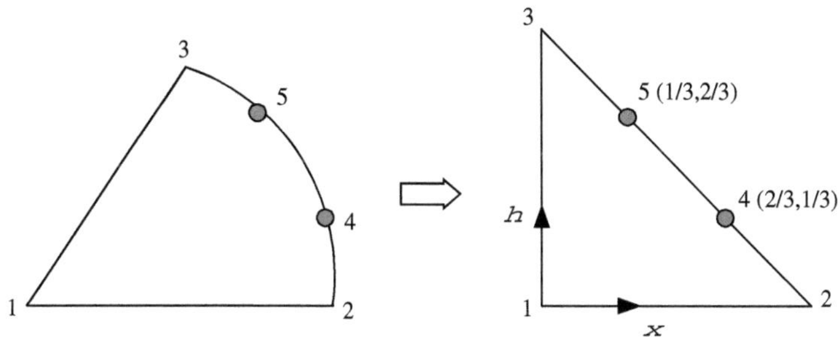

Fig. 4.8 The isoparametric transformation for the *curved* triangle

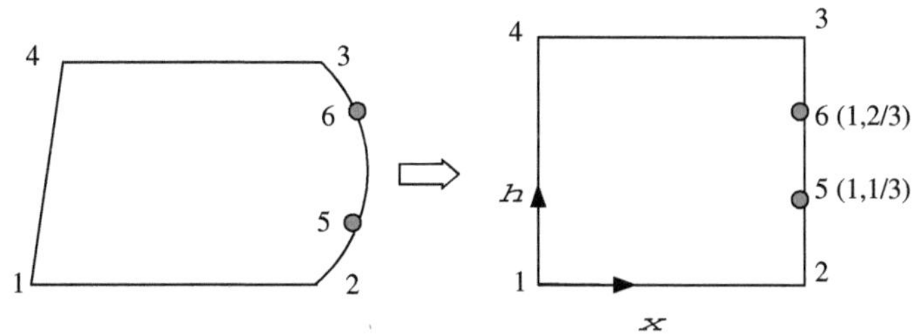

Fig. 4.9 The isoparametric transformation for the *curved* quadrilateral

$$
\begin{aligned}
\mathbf{x}(\xi,\eta) &= \sum_{k=0}^{4} v_k(\xi,\eta)\mathbf{x}_k,\\
v_0(\xi,\eta) &= 1-\xi-\eta,\\
v_1(\xi,\eta) &= \xi(1-3\eta)(1-1.5\eta),\\
v_2(\xi,\eta) &= \eta(1-4.5\xi\eta),\\
v_3(\xi,\eta) &= 9\xi\eta(1-1.5\eta),\\
v_4(\xi,\eta) &= 4.5\xi\eta(-1+3\eta).
\end{aligned}
$$

The isoparametric transformation for quadrilateral shown in Fig. 4.9 can be written as

$$
\begin{aligned}
\mathbf{x}(\xi,\eta) &= \sum_{k=1}^{6} v_k(\xi,\eta)\mathbf{x}_k,\\
v_1(\xi,\eta) &= (1-\xi)(1-\eta)\,,\\
v_2(\xi,\eta) &= \xi\left(1-\frac{11}{2}\eta+9\eta^2-\frac{9}{2}\eta^3\right),\\
v_3(\xi,\eta) &= \xi\eta\frac{1}{2}(1-3\eta)(2-3\eta),
\end{aligned}
$$

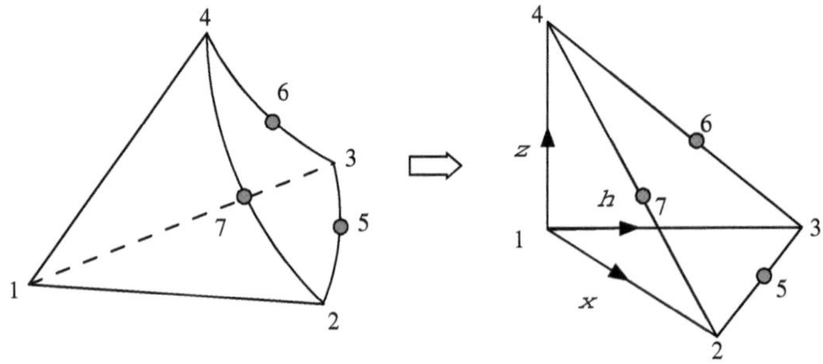

Fig. 4.10 The isoparametric transformation for the *curved* quadrilateral

$$\begin{aligned} v_4(\xi,\eta) &= \eta\,(1-\xi)\,, \\ v_5(\xi,\eta) &= \xi\eta\frac{9}{2}(1-\eta)(2-3\eta), \\ v_6(\xi,\eta) &= \xi\eta(-\frac{9}{2})(1-\eta)(1-3\eta). \end{aligned}$$

For the three-dimensional case, we give the isoparametric transformation for the curved tetrahedral and hexahedral grids using the second order polynomial approximation in Figs. 4.10 and 4.11. The transformation for the curved tetrahedron in Fig. 4.10 is

$$\begin{aligned} \mathbf{x} &= \sum_{k=1}^{6} v_k(\xi,\eta,\zeta)\mathbf{x}_k, \\ v_0(\xi,\eta,\zeta) &= 1-\xi-\eta-\zeta, \\ v_1(\xi,\eta,\zeta) &= \xi-2\xi(\eta+\zeta), \\ v_2(\xi,\eta,\zeta) &= \eta-2\eta(\xi+\zeta), \\ v_3(\xi,\eta,\zeta) &= \zeta-2\zeta(\xi+\eta), \\ v_4(\xi,\eta,\zeta) &= 4\xi\eta, \\ v_5(\xi,\eta,\zeta) &= 4\eta\zeta, \\ v_6(\xi,\eta,\zeta) &= 4\xi\zeta. \end{aligned}$$

And the transformation for the curved hexahedron in Fig. 4.11 is

$$\begin{aligned} \mathbf{x} &= \sum_{k=1}^{13} v_k(\xi,\eta,\zeta)\mathbf{x}_k, \\ v_1(\xi,\eta,\zeta) &= \xi(\xi-1)\eta(\eta-1)(1-\zeta)/8, \\ v_2(\xi,\eta,\zeta) &= \xi(\xi+1)\eta(\eta-1)(1-\zeta)/8, \end{aligned}$$

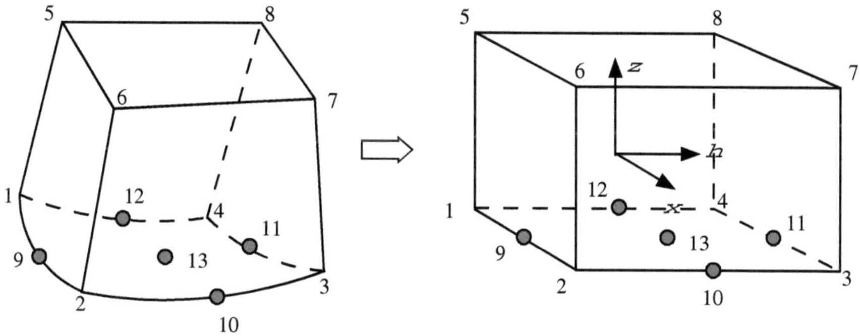

Fig. 4.11 The isoparametric transformation for the *curved* hexahedron

$$
\begin{aligned}
v_3(\xi,\eta,\zeta) &= \xi(\xi+1)\eta(\eta+1)(1-\zeta)/8,\\
v_4(\xi,\eta,\zeta) &= \xi(\xi-1)\eta(\eta+1)(1-\zeta)/8,\\
v_5(\xi,\eta,\zeta) &= (1-\xi)(1-\eta)(1+\zeta)/8,\\
v_6(\xi,\eta,\zeta) &= (1+\xi)(1-\eta)(1+\zeta)/8,\\
v_7(\xi,\eta,\zeta) &= (1+\xi)(1+\eta)(1+\zeta)/8,\\
v_8(\xi,\eta,\zeta) &= (1-\xi)(1+\eta)(1+\zeta)/8,\\
v_9(\xi,\eta,\zeta) &= (1-\xi^2)\eta(\eta-1)(1-\zeta)/4,\\
v_{10}(\xi,\eta,\zeta) &= \xi(\xi+1)(1-\eta^2)(1-\zeta)/4,\\
v_{11}(\xi,\eta,\zeta) &= (1-\xi^2)\eta(\eta+1)(1-\zeta)/4,\\
v_{12}(\xi,\eta,\zeta) &= \xi(\xi-1)(1-\eta^2)(1-\zeta)/4,\\
v_{13}(\xi,\eta,\zeta) &= (1-\xi^2)(1-\eta^2)(1-\zeta)/2.
\end{aligned}
$$

All the transformations proposed above can satisfy the condition (4.2). Transformations for other types of curved elements or higher order approximation can be also derived with similar methods.

4.2 The Nonoscillatory High-Order FVM in Mixed Grids

The high-order FVM on mixed grids is rarely found on the references. Tsoutsanis et al. [1] generalized the quadrature-free WENO FV methods proposed by Dumbser [2] to three-dimensional mixed elements. Whereas the stencils choosing method used by Tsoutsanis is extremely difficult and it destroys the quadrature-free property of Dumbser's WENO schemes, which is not suitable for the engineering applications. The purpose of this chapter is to apply the high-order limiters in Chap. 3 to FVM on mixed elements. And in the following sections, we can observe that these limiters on mixed elements still keep their original high-order accuracy and efficiently implemented properties, which are not possessed by other methods in the references by Hu and Shu [3], Dumbser et al. [2].

First the discretization for the high-order FVM (2.2) in the isoparametric element can be expressed as

$$\frac{\mathrm{d}}{\mathrm{d}t}\bar{U}_i + \frac{1}{\bar{\Omega}_i}\oint_{\partial\Omega^{\mathrm{iso}}} \mathbf{F}_n \cdot \mathbf{n}J_s\mathrm{d}S = 0, \quad \mathbf{F}_n = (F - F_v, G - G_v, H - H_v)$$

where Ω^{iso} is the isoparametric element, J_s is the Jacobian determinant on the surface. The flux can be computation as

$$\oint_{\partial\Omega_i} \mathbf{F}_n \cdot \mathbf{n}\mathrm{d}S = \sum_{k=1}^{N_F} \int_{\partial\Omega_{i,k}^{\mathrm{iso}}} \mathbf{F}_n \cdot \mathbf{n}J_s\mathrm{d}S = \sum_{k=1}^{N_F}\left(\sum_{\phi=1}^{N_G} \omega_g \hat{F}_n(U^R, U^L)\cdot \mathbf{n}S_k^{\mathrm{iso}}\right),$$

with the Gaussian points and weights ω_g easily obtained on the isoparametric element. Now it is necessary to evaluate the outward normal vectors or the outward normal area vectors $\mathbf{n}S_k^{\mathrm{iso}}$. Using the isoparametric transformation, we have

$$\mathbf{n}S_k^{\mathrm{iso}} = \pm\left(\frac{\mathrm{D}(y,z)}{\mathrm{D}(\xi,\eta)}, \frac{\mathrm{D}(z,x)}{\mathrm{D}(\xi,\eta)}, \frac{\mathrm{D}(x,y)}{\mathrm{D}(\xi,\eta)}\right)$$

where $\pm$ is determined by the outward normal direction of $\partial\Omega_{i,k}$. In the implementation, we do not store the Jacobian values on the Gaussian points but to compute these values directly in computational process. Since the complexity of the isoparametric transformation will affect the computational cost, our simplified transformation in Sect. 4.1 is beneficial to the computational time.

4.2.1 The k-Exact Reconstruction and Secondary Reconstruction

The reconstruction on the element i can also use (2.10)

$$u_i(\mathbf{x}) = \bar{u}_i + \sum_{l=1}^{K} u_i^l \varphi_{l,i}(\mathbf{x})$$

In Chap. 2 we choose the order-complete basis for the triangular and tetrahedral grids. A better choice for the quadrilateral and hexahedral grids is to use the tensor product generated by one- dimensional basis. However, to make the schemes more general on mixed elements, the order-complete basis (2.11) is used for all types of mixed elements. Equation (2.13) is used to reconstruct the polynomial and on mixed elements it can be written as

$$\bar{u}_i + \sum_{l=1}^{K} u_i^l \left(\frac{1}{\bar{\Omega}_j} \int_{\Omega^{\text{iso}}} \varphi_{l,i}(\mathbf{x}) J_j d\Omega \right) = \bar{u}_j, \; j \in S_i \tag{4.4}$$

where J_i is the Jacobian determinant for the isoparametric transformation on the element j. The left question is how to choose the reconstruction stencils S_i on the mixed elements. The main principle still follows the one used in Sect. 2.3, i.e.,namely conservation of the mean, k-exactness and compact support [4]. With these principles, we choose the face-neighbor cells as the first reconstruction stencils, and then take them as the outer border to choose face-neighbor cells as the second reconstruction stencils. The third and fourth reconstruction can be chosen in a similar way. This method works efficiently on triangular and tetrahedral grids, whereas the number of the stencils would be quite large on quadrilateral and hexahedral grids. So the following restriction would be added to the choosing principle. Since the $(k+1)$-th reconstruction step relies on the k-th reconstruction step, the reconstruction stencils for the $(k+1)$-th step is to choose the stencils sharing at least $(N_f - d)$ faces with the stencils at the k-th reconstruction step, where N_f is the faces for the central cell and d is the dimension. This restriction will make the reconstruction stencils more compact. Fig. 4.12 shows the reconstruction stencils for the fourth order reconstruction on the two-dimensional quadrilateral grids and the third order reconstruction on the three-dimensional hexahedral grids. Usually, the number of reconstruction stencils among 1.5 to 2 times DOFs of the polynomial can guarantee that no singularities occur in the reconstruction step.

To provide the reconstruction polynomial for the limiters, we still adopt the secondary efficient reconstruction. Since the basis function is the same, the secondary reconstruction in mixed elements is the same as the one in two-dimensional triangular, and three- dimensional tetrahedral grids, i.e., the formula (3.13) and (3.14). The stencils for the secondary reconstruction is different in every limiting strategy, thus we give the choosing method for the limiting stencils on mixed elements below.

4.2.2 The k-Exact WENO Limiter

Note the stencils for the secondary reconstruction as T_i. For $j \in T_i$, we still use the three principles in choosing the limiting stencils, i.e.,

(1) Cell j in Eqs. (3.13) and (3.14) belong to S_i. The condition guarantees the accuracy of the secondary reconstruction.
(2) The stencils should be symmetrical and the farthest away from the central cell topologically.
(3) The stencils should contain every direction with respect to the central cell.

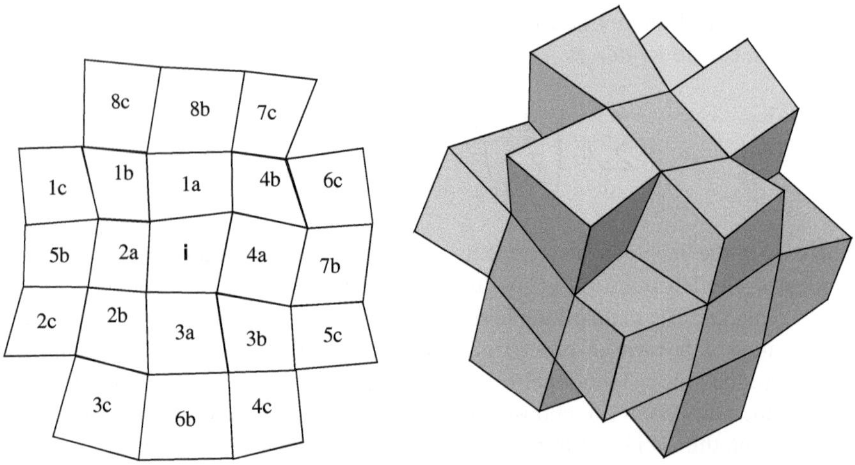

Fig. 4.12 Reconstruction stencils for mixed elements. *Left* Fourth order reconstruction for quadrilateral grids. *Right* Third order for hexahedral grids

Table 4.1 The number of shared vertex with the central cell for the secondary reconstruction in the k-exact WENO limiter

Central element type	Triangle	Quadrilateral	Tetrahedron	Prism	Hexahedron
3rd order SR	1	1	2	2	2
4th order SR	0	0	1	1	1

According to these three principles, we can also give the limiting stencils for the k-exact WENO limiters on mixed elements. For the third and fourth order reconstructions on mixed elements, the reconstruction stencils can be obtained using the number of shared vertex with the central cell. The choosing rule is shown in Table 4.1. Figure 4.13 shows the limiting stencils for the fourth order reconstruction in quadrilateral, which share no vertex with the central cell. For the schemes of fourth order and below, this choosing method contains at least one smooth stencil and the stencils number is small. For higher order schemes, the three principles are still valid but the difficulty may be a bit large for we need the topological distance to the central cell.

With the limiting stencils T_i, the secondary reconstruction is carried out for $j \in T_i$ and the candidate polynomial $u_{j \to i}(\mathbf{x})$ using the formula (3.13) and (3.14). The WENO weighted procedure is still the same as Sect. 3.3.2 as long as the reconstruction basis is the same. This also indicates the k-exact WENO scheme is still simple and efficient in mixed elements.

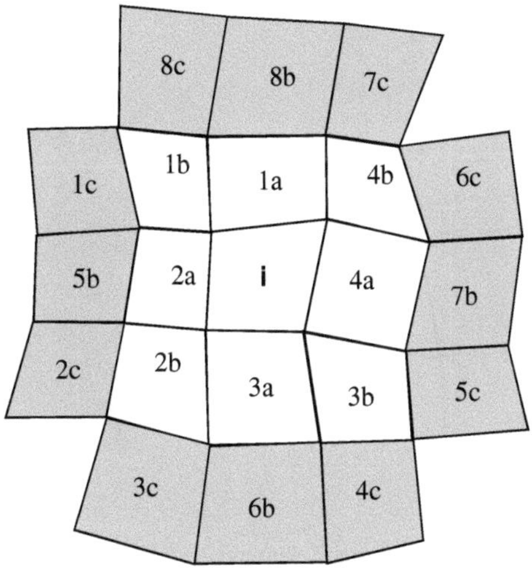

Fig. 4.13 The secondary reconstruction stencils in k-exact WENO limiters on quadrilateral (*shadow marked*)

4.2.3 The WBAP Limiter Based on the Successive Limiting Procedure

In Sect. 3.4, a summary about the merits of WBAP is given, i.e., compactness and genuinely quadrature-free. The two merits make the WBAP work quite efficiently on mixed elements. Figure 4.14 shows the limiting stencils for WBAP limiters on quadrilateral and hexahedron. Only the face-neighbor cells are used, e.g., $T_i = \{1a, 2a, 3a, 4a\}$ on quadrilateral. The left process in identical to that in Sect. 3.4, which also indicates the flexibility of WBAP limiters on mixed grids.

Moreover, the group weighted limiter can also be generalized to mixed elements using similar methods as WBAP limiters.

4.3 The Curved Boundary Treatment

In the last section, we transform the mixed grids to the regular region and carry out the quadrature over these regular regions. This method also works in treating the curved boundary. The premise for the isoparametric transformation is enough points to represent the curve surface. Whereas the grid generation software only provides the coordinates of the vertex. Thus, we need to reconstruct the curved surface. For the two-dimensional case shown in Fig. 4.15, the reconstruction of the curved line $\widetilde{12}$ between vertex 1 and 2 requires points on the left- and right- hand-sides, i.e., vertex 0 and 3. Then a third-order polynomial is constructed for the curved line $\widetilde{12}$. We can compute the coordinates for the additional control points for Figs. 4.8 and 4.9

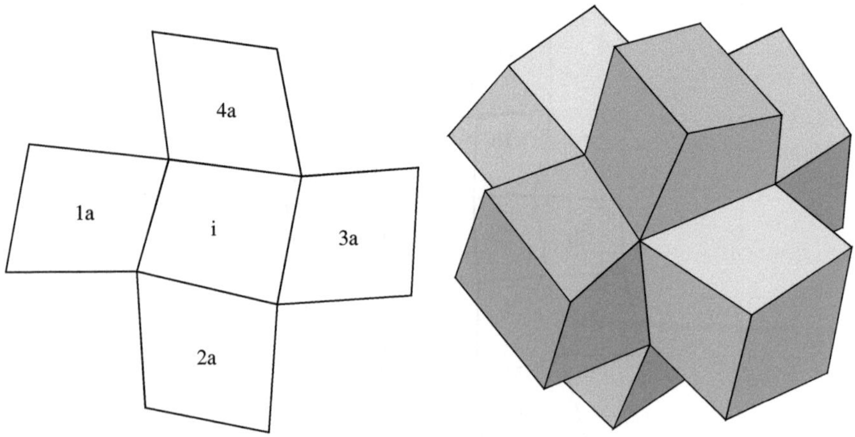

Fig. 4.14 Reconstruction stencils for mixed elements. *Left* Fourth order reconstruction for quadrilateral grids. *Right* Third order for hexahedral grids

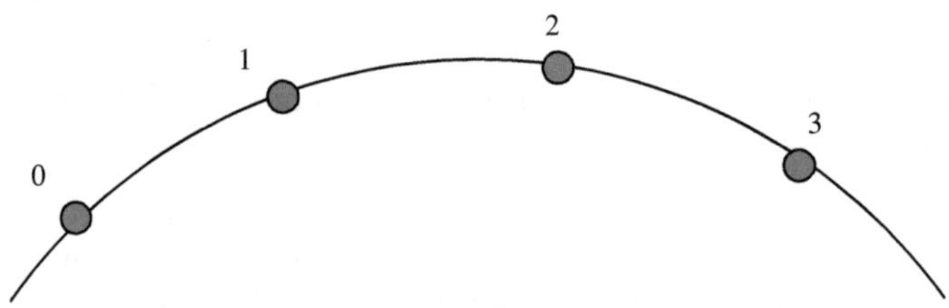

Fig. 4.15 The representation for the *curved line*

using the third-order polynomial. For the three-dimensional cases, the best way to give the coordinates of the additional control points in Figs. 4.10 and 4.11 is using the real geometry. The reconstruction of three-dimensional curved surface is still not so satisfying. More detail information could be found in [5].

With isoparametric transformation for the curved surface and element, the integration on face and volume can be carried out on the regular isoparametric elements. For an arbitrary function $f(x, y, z)$, the quadrature would be

$$\int_{\partial\Omega} f(x,y,z)\cdot \mathbf{n}\mathrm{d}S = \pm \int_{\partial\Omega^{\mathrm{iso}}} f(\xi,\eta)\cdot\left(\left|\frac{\mathrm{D}(y,z)}{\mathrm{D}(\xi,\eta)}\right|, \left|\frac{\mathrm{D}(z,x)}{\mathrm{D}(\xi,\eta)}\right|, \left|\frac{\mathrm{D}(x,y)}{\mathrm{D}(\xi,\eta)}\right|\right)\mathrm{d}\xi\mathrm{d}\eta$$
$$\int_{\Omega} f(x,y,z)\mathrm{d}x\mathrm{d}y\mathrm{d}z = \int_{\Omega^{\mathrm{iso}}} f(\xi,\eta,\zeta)\left|\frac{\mathrm{D}(x,y,z)}{\mathrm{D}(\xi,\eta,\zeta)}\right|\mathrm{d}\xi\mathrm{d}\eta\mathrm{d}\zeta$$

where $|\cdot|$ denotes the Jacobian determinant. Since the face integration is needed in the flux evaluation in Eq. (2.4), $\pm$ depends on the external normal direction of $\partial\Omega$. After this operation, other calculations on the curved face and cell are the same as the straight ones.

4.4 Numerical Tests

This section is going to verify the high-order accuracy and nonoscillation properties for finite volume schemes on the mixed grids and curved boundaries. In the meanwhile, some complicated flow problems are computed to validate the efficiency of the high-order FVM.

4.4.1 Supersonic Vortex

This case is chosen to validate the accuracy of the schemes for solving problem with curved boundary. We consider an isentropic supersonic flow between two concentric circular arcs of radii $r_i = 1$ and $r_o = 1.384$ in the first quadrant, which can be found in [6]. The exact density in terms of radius r is given by

$$\rho = \rho_i \left(1 + \frac{\gamma - 1}{2} M_i^2 \left(1 - \left(\frac{r_i}{r}\right)^2\right)\right)^{1/(\gamma-1)}$$

The velocity and pressure are given by

$$\| \mathbf{v} \| = \frac{c_i M_i}{r}, \quad P = \frac{\rho^\gamma}{\gamma}$$

where c_i is the speed of sound on the inner circle. The Mach number on the inner circle M_i is set to 2.25 and the density to one.

The problem is solved on a series of refined triangular unstructured grids consisting of 464, 1,894, 7,524, and 29,990 cells. The two coarsest grids are shown in Fig. 4.16. In each simulation we compute the solution until the residual comes to machine zero. The errors in terms of the density are shown in Table 4.2. For the k-exact WENO schemes, the constant λ_0 is set to be 100. We can see that the fourth order unlimited FV scheme with straight wall representation gets low grid convergence results, while the unlimited FV and k-exact WENO schemes using curved boundary representation can reach the desired order in terms of L_1 error, and the order of L_∞ is one order lower than the L_1 error. That is due to the fact that the stencils for high-order scheme on the boundary cells are chosen in a biased way.

4.4.2 The Supersonic Viscous Flow Around the NACA0012 Airfoil

This test case solves the viscous flow with shock waves using the high-order FVM. The supersonic flow at Mach 2 flows around the NACA0012 airfoil. The Reynolds number based on the chord is 10,000, and the angle of attach is 2°. The flow is

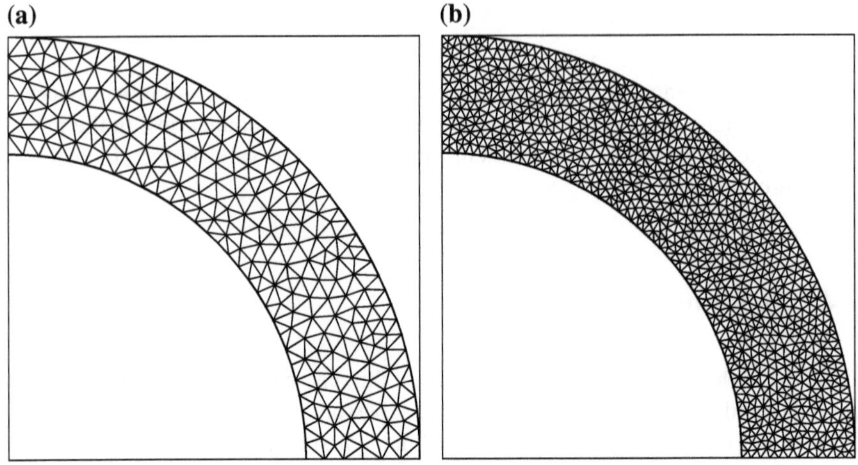

Fig. 4.16 Grids for the accuracy test in supersonic vortex, $h_0 = 0.384$. **a** $h_0/6$ **b** $h_0/12$

Table 4.2 Accuracy test for supersonic vortex using fourth order FV scheme, h_0=0.384

Case	Grid	L_1 error	Order	L_∞ error	Order
Unlimit,straight	$h_0/6$	3.49E-04		4.51E-03	
	$h_0/12$	4.24E-05	3.04	9.15E-04	2.30
	$h_0/24$	5.97E-06	2.83	2.69E-04	1.76
	$h_0/48$	1.11E-06	2.43	7.81E-05	1.79
Unlimit,curve	$h_0/6$	2.48E-04		2.97E-03	
	$h_0/12$	1.74E-05	3.83	3.77E-04	2.98
	$h_0/24$	1.12E-06	3.95	4.51E-05	3.07
	$h_0/48$	8.32E-08	3.75	5.30E-06	3.09
k-Exact WENO,curve	$h_0/6$	2.27E-04		2.79E-03	
	$h_0/12$	1.57E-05	3.86	3.41E-04	3.03
	$h_0/24$	1.01E-06	3.96	3.95E-05	3.11
	$h_0/48$	7.28E-08	3.79	4.48E-06	3.14

calculated using 9,378 mixed triangular and quadrilateral grids and 5,821 grids points shown in Fig. 4.17. The second to fourth order FV schemes are used with the WBAP limiter. The Riemann solver is the Roe scheme.

The Mach number contour of this supersonic flow is shown on the left of Fig. 4.18, where we can observe the detached shock near the head. From the convergent line shown on the right of Fig. 4.18, the WBAP limiters exhibits quite satisfied convergent abilities with second to fourth order schemes converging to machine zero. The pressure coefficients and skin friction coefficients on the surface are shown in Fig. 4.19, where the reference values are the results of second order scheme on refined grids. We observe that all schemes give nearly the same results for pressure coefficients and third and fourth order schemes can come closer to reference values than sec-

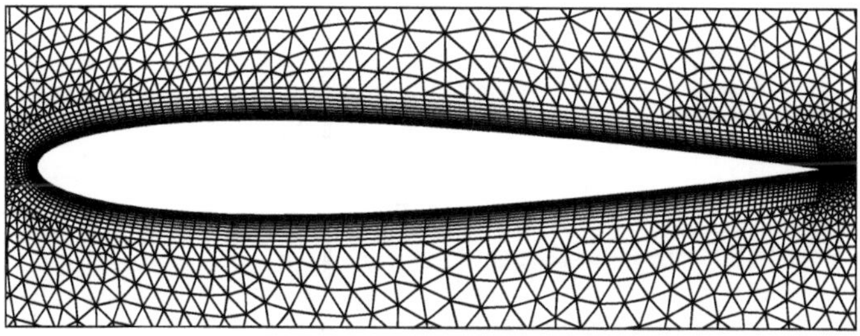

Fig. 4.17 The mixed triangular and quadrilateral grids for NACA0012 airfoil

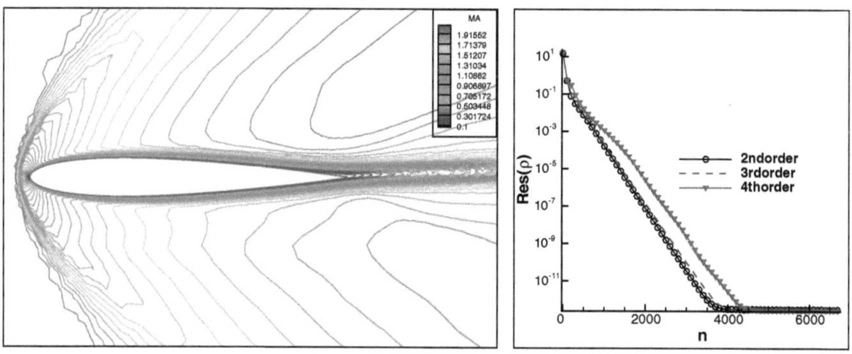

Fig. 4.18 *Left* is the contour of Mach number for the supersonic flow around NACA0012. *Right* is the convergence history for different orders

ond order scheme. Whereas for the skin friction coefficients, the superiority of higher order schemes are obvious especially on the leading edge of airfoil. From the figures, the results are already order convergent. This case also implies that the drag-related parameters in viscous flows require higher for the accuracy of the scheme.

4.4.3 The Three-Dimensional Constructed Solutions

The description of the problem is seen in Sect. 2.5.1. This section will validate the accuracy of high-order FVM on different types of grids. As shown in Fig. 4.20, the test grids are the hexahedral prismatic and mixed hexahedral and prismatic grids.

The second to fourth order FV schemes are used for the accuracy test on these three types of grids and the results are shown in Tables 4.3 and 4.4. All results can achieve the expected orders and the high-order schemes are much more accurate than low- order ones. Even the WBAP limiter is used on the mixed grids, the scheme still keeps high-order accuracy.

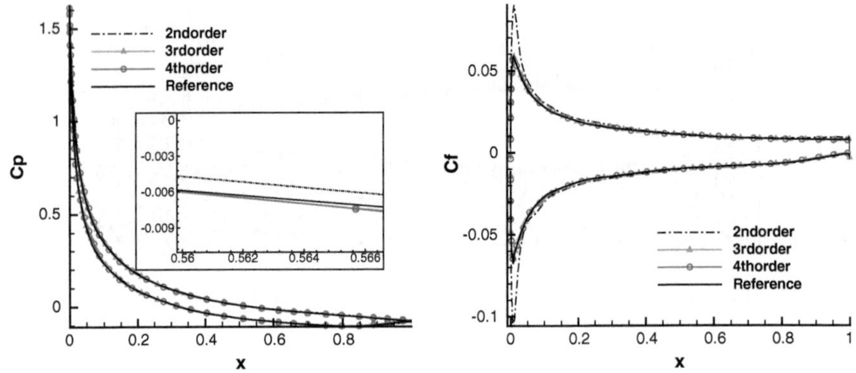

Fig. 4.19 The distributions for the surface pressure (*left*) and skin friction coefficients (*right*)

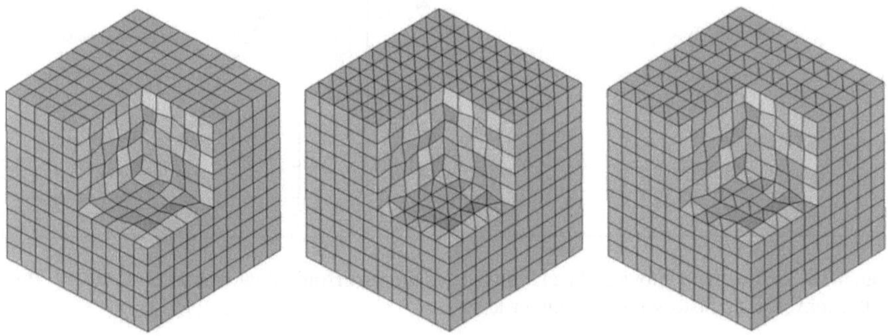

Fig. 4.20 The three-dimensional computation grids. For *left* to *right* are hexahedron, prism, and mixed grids, respectively

4.4.4 Lax Shock Tube Problem

This test case is chosen to test the capability of k-exact WENO schemes in capturing the strong shock wave and contact discontinuity. The problem is one-dimensional in nature, and the initial conditions are

$$(\rho, u, v, w, p)(\mathbf{x}) = \begin{cases} (\rho_L, u_L, 0, 0, p_L) & \text{if } x \leq 0.5 \\ (\rho_R, u_R, 0, 0, p_R) & \text{if } x > 0.5 \end{cases}$$

We consider the initial conditions given by

$$(\rho_L, u_L, p_L) = (0.445, 0.698, 3.528),$$
$$(\rho_R, u_R, p_R) = (0.5, 0, 0.571)$$

Table 4.3 The accuracy test on three-dimensional hexahedron and prism grids using FVM

Scheme	Grid	Hexahedron grid				Prism grid			
		$L1$ error	Order	L error	Order	$L1$ error	Order	L error	Order
2nd order	1/10	7.76E-02		1.65E-01		9.64E-02		1.90E-01	
	1/20	2.03E-02	1.93	4.56E-02	1.85	2.55E-02	1.92	6.66E-02	1.51
	1/30	1.12E-02	1.47	2.01E-02	2.02	1.39E-02	1.49	2.69E-02	2.23
	1/40	6.80E-03	1.74	1.09E-02	2.12	8.41E-03	1.76	1.65E-02	1.71
3rd order	1/10	5.64E-02		0.148269		5.14E-02		0.117487	
	1/20	7.92E-03	2.83	2.07E-02	2.84	7.03E-03	2.87	1.59E-02	2.89
	1/30	2.39E-03	2.96	6.16E-03	2.99	2.10E-03	2.97	4.75E-03	2.98
	1/40	1.01E-03	2.98	2.58E-03	3.02	8.89E-04	3	2.00E-03	3.01
4th order	1/10	1.55E-02		2.71E-02		1.23E-02		4.32E-02	
	1/20	9.46E-04	4.03	1.65E-03	4.04	6.97E-04	4.14	2.80E-03	3.95
	1/30	1.82E-04	4.06	3.19E-04	4.05	1.32E-04	4.11	5.25E-04	4.13
	1/40	5.68E-05	4.05	1.00E-04	4.03	4.10E-05	4.05	1.67E-04	3.98

Table 4.4 The accuracy comparison between unlimited FV and FV with WBAP limiters on mixed hexahedron and prism grids

Scheme	Grid	Unlimit				WBAP limiter			
		$L1$ error	Order	L error	Order	$L1$ error	Order	L error	Order
2nd order	1/10	2.98E-02		9.37E-02		8.80E-02		0.185076	
	1/20	7.55E-03	1.98	3.75E-02	1.32	2.23E-02	1.98	5.63E-02	1.72
	1/30	3.35E-03	2.01	2.41E-02	1.09	1.21E-02	1.52	2.33E-02	2.18
	1/40	1.87E-03	2.01	1.68E-02	1.26	7.22E-03	1.78	1.34E-02	1.93
3rd order	1/10	5.35E-02		0.141037		6.43E-02		0.149113	
	1/20	7.41E-03	2.85	1.96E-02	2.85	7.62E-03	3.08	2.03E-02	2.88
	1/30	2.22E-03	2.97	5.82E-03	2.99	2.16E-03	3.11	6.20E-03	2.92
	1/40	9.41E-04	2.99	2.44E-03	3.02	8.84E-04	3.1	2.66E-03	2.95
4th order	1/10	1.53E-02		2.85E-02		1.84E-02		6.01E-02	
	1/20	8.69E-04	4.13	1.86E-03	3.94	7.65E-04	4.59	4.32E-03	3.8
	1/30	1.64E-04	4.11	3.60E-04	4.05	1.67E-04	3.75	1.01E-03	3.58
	1/40	5.13E-05	4.04	1.12E-04	4.06	5.14E-05	4.1	2.95E-04	4.28

which is the Harten-Lax problem. The domain is taken as $[0, 1] \times [0, 0.1] \times [0, 0.1]$, and the grids for the computation are shown in Fig. 4.21 with the grid size $h = 0.01$ and $h = 0.005$ for the tetrahedral and hexahedral grids, respectively. The problems are computed using the k-exact WENO limiter with $\lambda_0 = 100$ and WBAP limiter with $\lambda_0 = 10$. From Figs. 4.22 and 4.23 we can clearly see that both the third and fourth order k-exact WENO and WBAP limiters can capture the discontinuity of the Harten-Lax tube without oscillations.

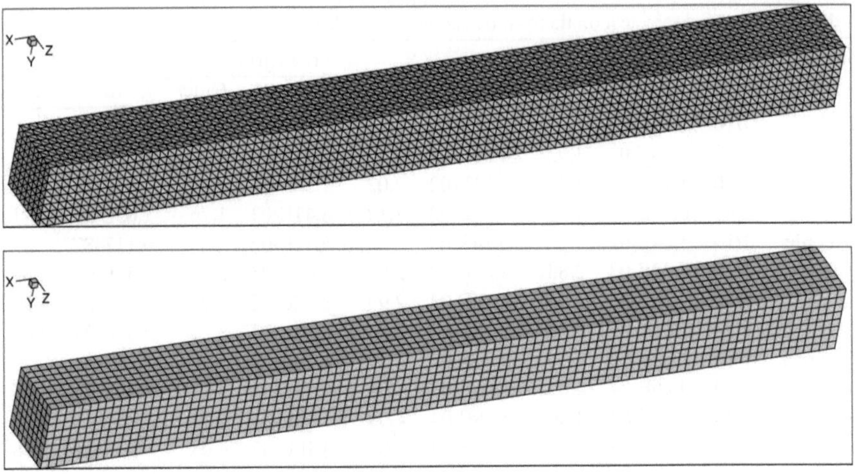

Fig. 4.21 Uniform tetrahedral and hexahedral girds in three dimensions

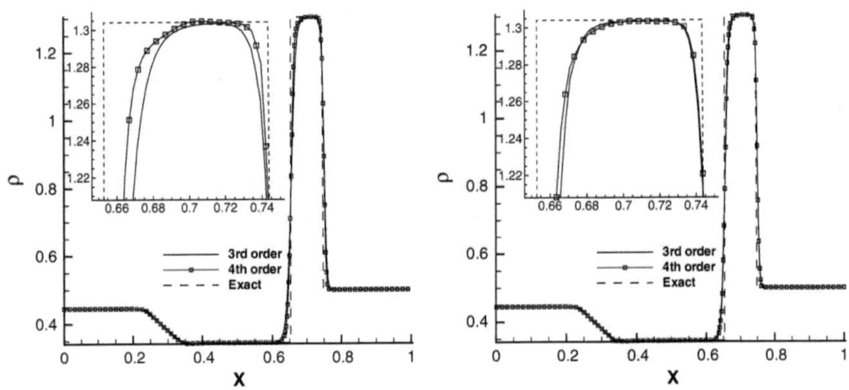

Fig. 4.22 Comparison of density distribution at for shock tube problems using k-exact WENO limiters. *Left* tetrahedral grids; *right* hexahedral grids

4.4.5 Transonic Flow Around ONERA M6 Wing

A transonic flow over the ONERA M6 wing geometry is considered in this test case. The M6 wing has a leading edge sweep angle of 30°, an aspect of 3.8, and a taper ratio of 0.562. The airfoil section of the wing is the ONERA "D" airfoil, which is a 10 % maximum thickness-to-chord ratio conventional section. The flow solutions are presented at a Mach number of 0.84 and an angle of attach of 3.06° using the second and third order FV method on a coarse mesh and a fine mesh. The coarse mesh contains 136,705 elements, 25,616 points, and 5,017 boundary points, and the

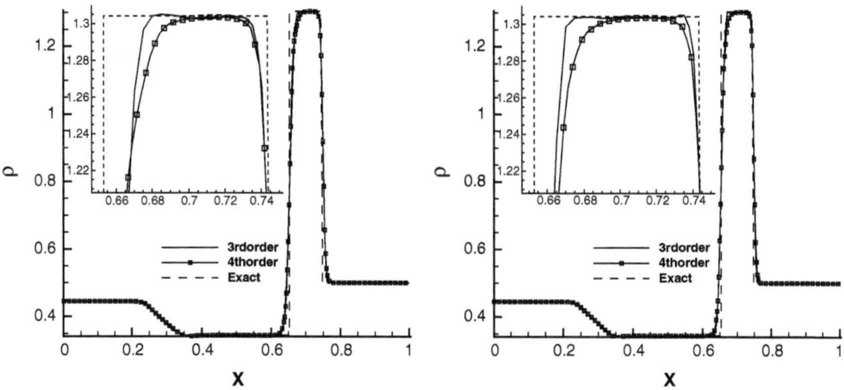

Fig. 4.23 Comparison of density distribution at for shock tube problems using WBAP limiters. *Left* tetrahedral grids; *right* hexahedral grids

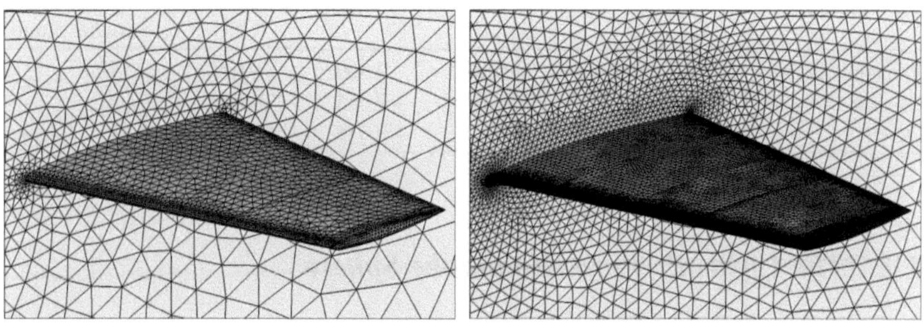

Fig. 4.24 Two sets of ONERA M6 grids. *Left* coarse grids; *right* fine grids

fine one 1,092,270 elements, 192,727 points, and 20,066 boundary points, which are both shown in Fig. 4.24.

Figure 4.25 shows the computed Mach number contours on the wing surface obtained using the third order k-exact WENO scheme on both grids, where we can observe near the shock wave there are no obvious oscillations. The computed pressure coefficient and entropy production distributions obtained by the second and third order k-exact WENO scheme are compared at six different span-wise stations in Figs. 4.26 and 4.27. From the pressure distribution the computed results can match the experimental one to some extent, and third order scheme on coarse grids can get a bit closer to finer grids' results than second order scheme but not very obvious. Whereas from the entropy production distributions, we can clearly find that higher order schemes can produce much less entropy than the lower order one whatever on coarse and fine grids.

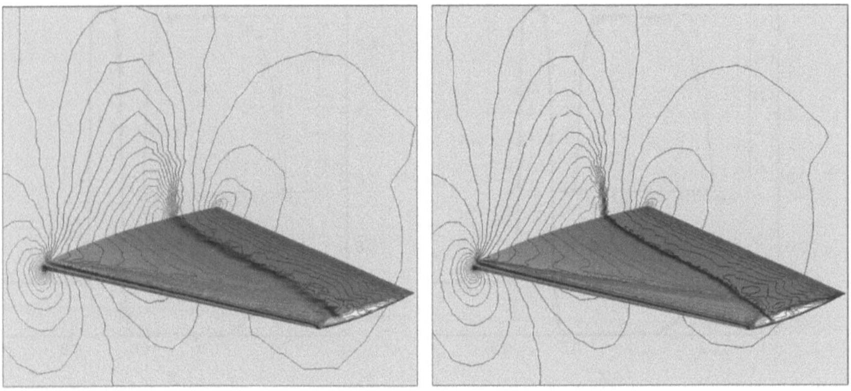

Fig. 4.25 Mach number contour using third order k-exact WENO scheme. *Left* coarse grids; *right* fine grids

Table 4.5 The comparison of lift and drag coefficients Cl, Cd, and Strouhal number

	Cd	Δ Cd	Cl	Δ Cl	Strouhal number
Johnson and Patel [7]	0.656	0.0035	−0.069	0.016	0.137
Gassner [8]	0.673	0.0031	−0.065	0.015	0.135
Ours	0.674	0.003	−0.055	0.013	0.133

4.4.6 The Unsteady Flow Around the Sphere

This test case is used to verify the capability of the cured boundary treatment in viscous flow. The unsteady flow around the sphere is chosen with Reynolds number Re = 300 and inlet Mach number $M_\infty = 0.3$. The radius of the sphere is 0.5 and the regions is chosen to the union of half sphere with radius $R = 10$ and cylinder with the same radius and height 40. Tetrahedra are used as the control volume with grid size $h = 0.014$ near the sphere and the cross-section is shown in Fig. 4.28. Totally, we have 2,065,612 elements and 347,203 grid points.

The third order FVM is used in the computation and curved boundary treatment is used on the wall boundary. The Q-criterion is used to identify the vortex around the sphere and the vortex is shown in Fig. 4.29. The Mach number contour at $z = 0$ section is shown in Fig. 4.30 where the wake is clearly seem. We draw the history for the lift and drag coefficients in Fig. 4.31 with the comparison between the results in this section and other references in Table 4.5. We can see the computed drag coefficient and Strouhal number is close to that in the reference and the average lift coefficient is closer to zero than other references.

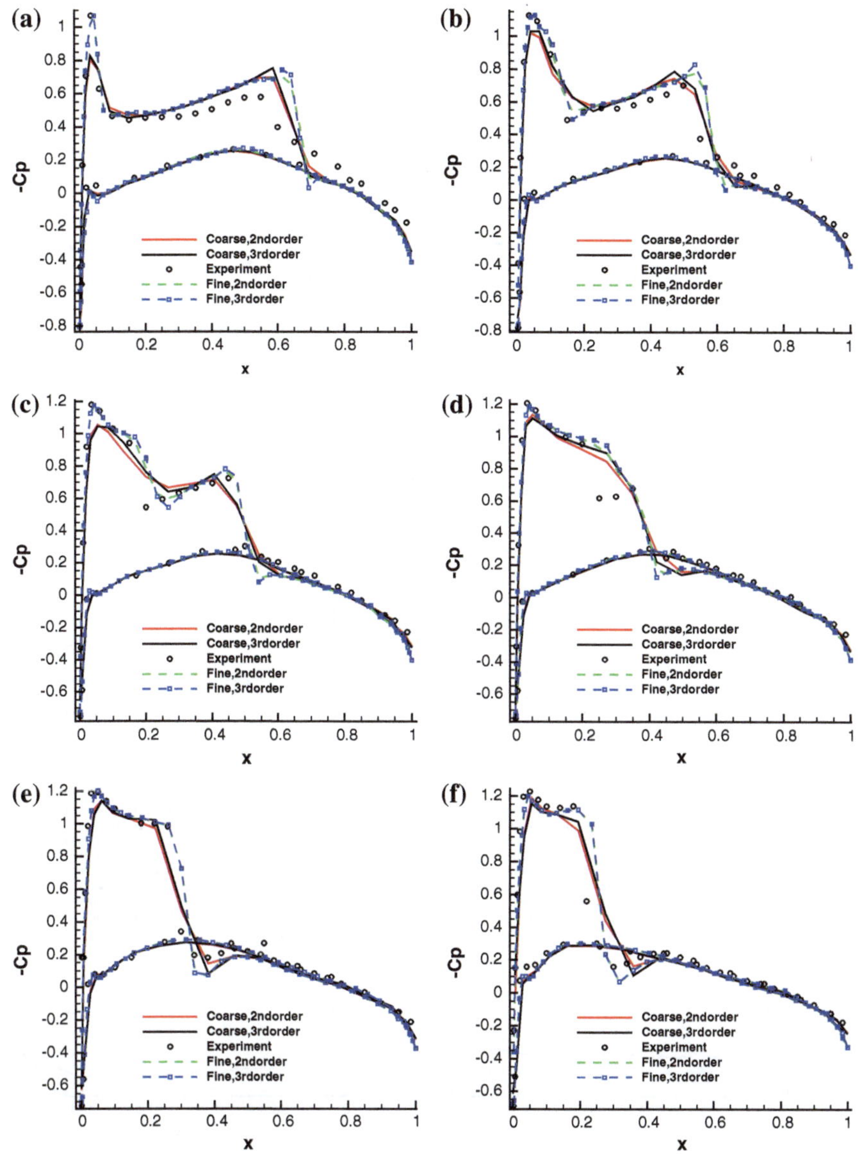

Fig. 4.26 Comparison of computed pressure coefficient distributions for wing section at different semi-span locations obtained by second and third order k-exact WENO scheme on coarse and fine grids for transonic flow past an ONERA wing at $M_\infty = 0.84$, $\alpha = 3.06°$. **a** 20 % **b** 44 % **c** 65 % **d** 80 % **e** 90 % **f** 95 %

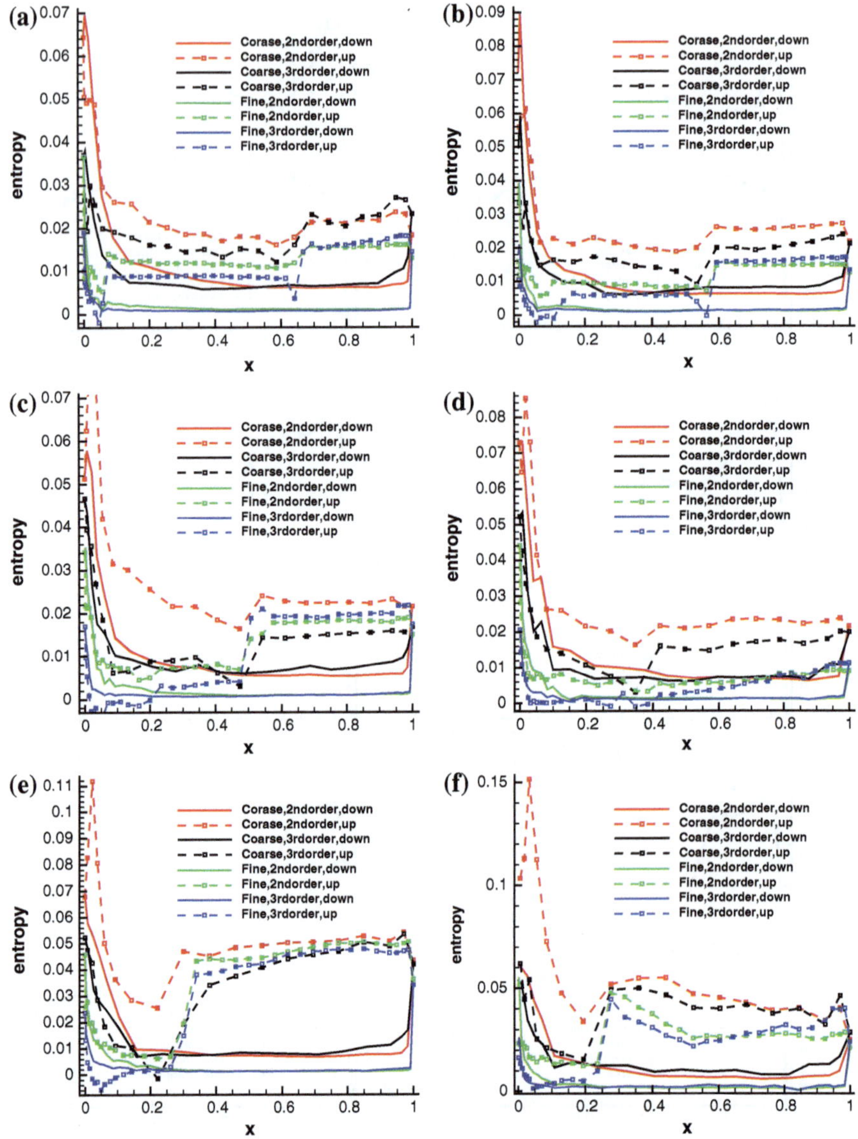

Fig. 4.27 Comparison of entropy production distributions for wing section at different semi-span locations obtained by second and third order k-exact WENO scheme on coarse and fine grids. **a** 20 % **b** 44 % **c** 65 % **d** 80 % **e** 90 % **f** 95 %

4.4.7 The Supersonic Flow Around the Sphere

This section computes the supersonic flow around the sphere to validate the curved boundary treatment and shock capturing ability in high-order FVM. The solutions are

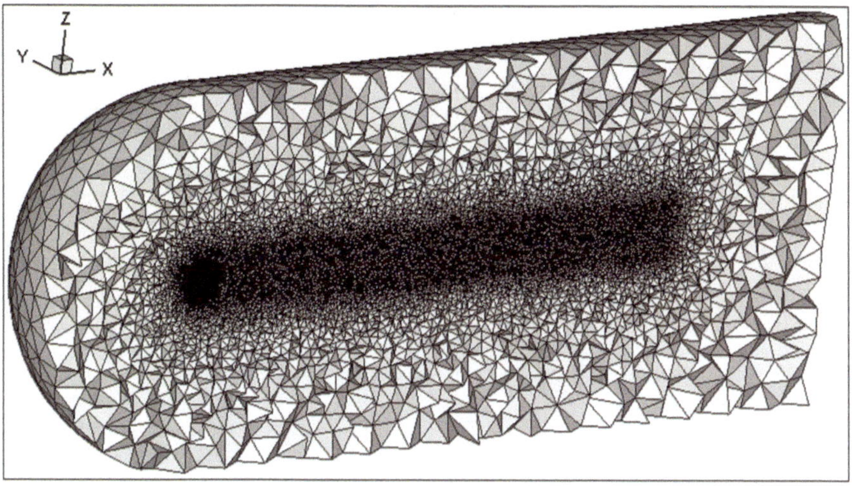

Fig. 4.28 The computational grids for the flow around the sphere

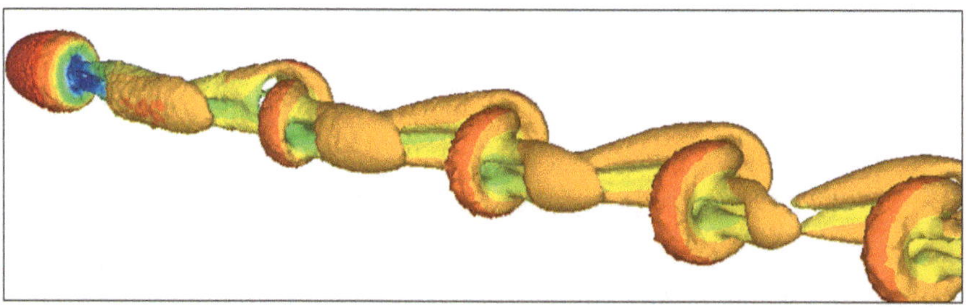

Fig. 4.29 The vortex using Q-criteria at $Q = 0.001$ colored by Mach number

presented at Mach number 2.0, Reynolds number 300 using the third and fourth order schemes. The Roe Riemann solver and WBAP limiters are used in the calculation. The mesh is the same as that in Sect. 2.6.5 with 458,915 tetrahedra and 77,844 grid points. The Mach number contour and skin pressure coefficients computed using the fourth order FV scheme are shown in Fig. 4.32, where we clearly observe the detached shock wave at the head and separations at the wake. Both third and fourth order schemes give the same total drag coefficient $Cd = 1.46$. Also the skin friction coefficients comparison shown in Fig. 4.33 indicate that the results are order convergent.

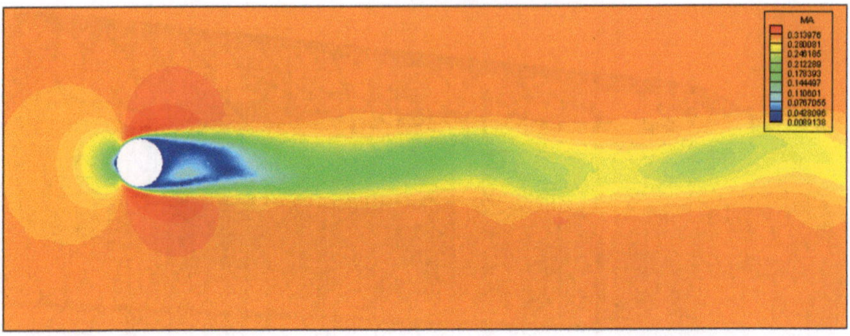

Fig. 4.30 The contour of Mach number in the wake at the section $z = 0$

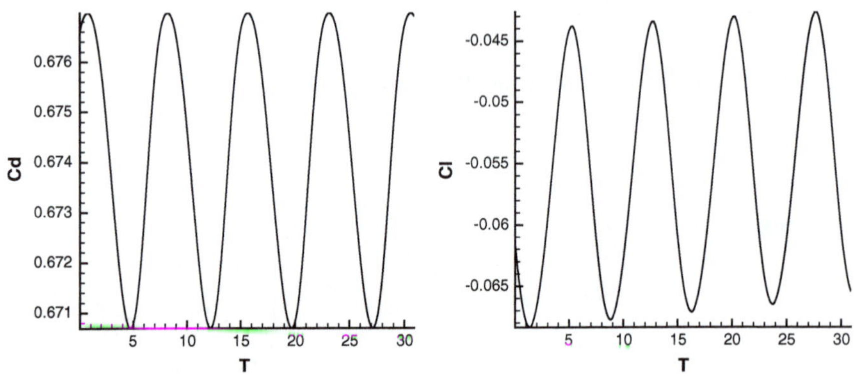

Fig. 4.31 The history of lift (*left*) and drag (*right*) coefficients

4.5 Conclusions

This chapter generalizes the nonoscillatory finite volume scheme to the mixed element and curved boundaries cases. First is to change the mixed and curved elements to the regular elements using isoparametric transformation. Thus, the high-order face and volume integrals can be carried out on the regular regions. Second, is to apply the high-order FVM on mixed unstructured grids through the solving of stencils choosing problem of k-exact reconstruction and the implementation of high-order limiters. In fact the high-order limiters proposed in Chap. 3 still work on mixed grids in a similar way as on the triangular or tetrahedral grids. On the mixed elements, the k-exact WENO limiters can also achieve quadrature-free and the WBAP limiters still remain compact. These also validate that the proposed limiters are easily implemented and computationally efficient. The curved boundary treatment mainly

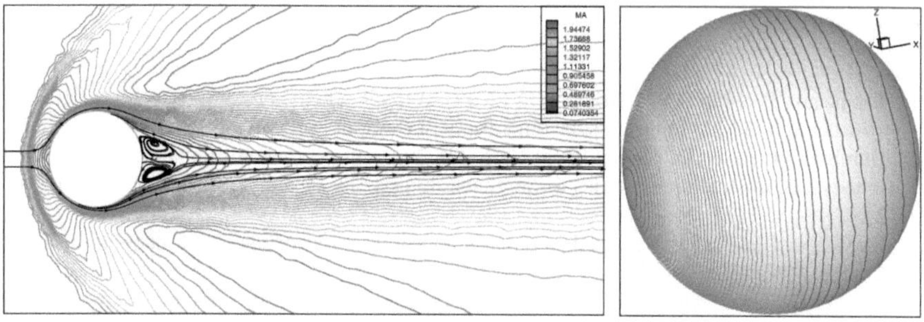

Fig. 4.32 The Mach number contour at cross-section $z = 0$ (*left*) and the pressure distribution (*right*) computed using the fourth order FV scheme

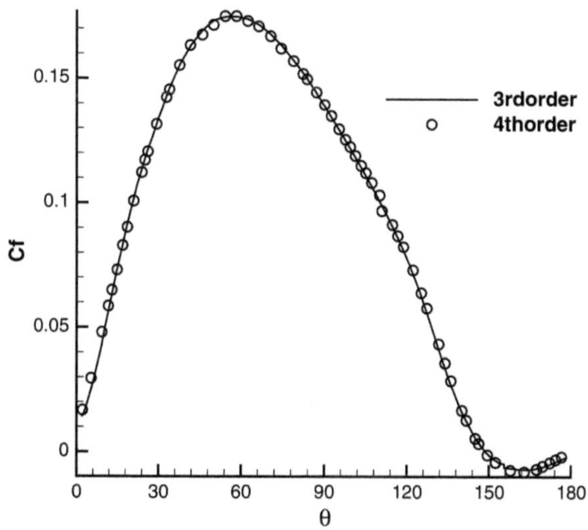

Fig. 4.33 The comparison of skin friction coefficients C_f between third and fourth order schemes

takes use of the current methods to construct the curved surface, and transform the mixed elements to regular elements in order to carry out the high- order quadrature on the face and volume. Finally, abundance of numerical examples are computed using the high-order FVM on mixed grids to validate the nonoscillatory property and to demonstrate the superiority of high-order schemes in resolving the boundary layer and shock waves in viscous flows.

References

1. Tsoutsanis P, Titarev VA, Drikakis D (2011) WENO schemes on arbitrary mixed-element unstructured meshes in three space dimensions. J Comput Phys 230:1585–1601
2. Dumbser M, Käser M, Titarev VA et al (2007) Quadrature-free non-oscillatory finite volume schemes on unstructured meshes for nonlinear hyperbolic systems. J Comput Phys 226:204–243
3. Hu C, Shu CW (1999) Weighted essentially non-oscillatory schemes on triangular meshes. J Comput Phys 150:97–127
4. Ollivier-Gooch C, Altena MV (2002) A high-order-accurate unstructured mesh finite-volume scheme for the advection-diffusion equation. J Comput Phys 181:729–752
5. Lübon C, Wagner S (2007) Three-dimensional discontinuous Galerkin codes to simulate viscous flow by spatial discretization of high order and curved elements on unstructured grids. New Results Numer Exp Fluid Mech 96:145–153
6. Krivodonova L, Berger M (2006) High-order accurate implementation of solid wall boundary conditions in curved geometries. J Comput Phys 211:492–512
7. Johnson TA, Patel VC (1999) Flow past a sphere up to a Reynolds number of 300. J Fluid Mech 378:19C70
8. Gassner G (2009) Discontinuous Galerkin method for the unsteady compressible Navier-Stokes equations. PhD thesis, University Stuttgart, Germany

Chapter 5
The Discontinuous Galerkin Method

Accuracy preserving and nonoscillatory shock capturing technique is also one of the bottlenecks in the development of discontinuous Galerkin method. In this chapter, a new limiter based on the secondary reconstruction and the WENO approach in the characteristic space is developed for the discontinuous Galerkin method. Specifically, the efficient secondary reconstruction technique is adopted to provide the candidate polynomials used in the weighted average procedure of the WENO approach. The secondary reconstructions are performed only on the face neighboring cells to keep the compactness of the discontinuous Galerkin method. Moreover, an improved version of the characteristic limiting procedure is proposed for the nonlinear Euler equations, which is considerably more efficient than the traditional characteristic limiting procedures. The resulting schemes are easy to implement and effective in capturing the shock waves. Some standard cases are computed to validate the accuracy and robustness of the proposed limiters for the DG methods.

5.1 Framework of the Discontinuous Galerkin Methods

This section describes the discretization of the discontinuous Galerkin schemes on unstructured grids. The equations are the Euler equations governing the unsteady compressible inviscid flow, which are written as

$$\frac{\partial U}{\partial t} + \frac{\partial F}{\partial x} + \frac{\partial G}{\partial y} = 0, \tag{5.1}$$

where U is the vector of the conservative variables given as $U = (\rho, \rho u, \rho v, \rho E)^T$. The detailed formulations of the flux terms are well known and are omitted here for brevity.

The discretization of Eq. (5.1) using the discontinuous Galerkin method is in terms of the weak formulation of Eq. (5.1),

W. Li, *Efficient Implementation of High-Order Accurate Numerical Methods on Unstructured Grids*, Springer Theses, DOI: 10.1007/978-3-662-43432-1_5,

$$\int_{\Omega_i} \frac{\partial U}{\partial t} W d\Omega + \oint_{\partial \Omega_i} \widehat{F}_n \cdot \mathbf{n} W ds - \int_{\Omega_i} \mathbf{F}_n \cdot \nabla W d\Omega = 0, \ \mathbf{F}_n = (F, G), \tag{5.2}$$

where $\partial \Omega_i$ denotes the boundary of Ω_i, $\mathbf{n}$ is the unit outward normal vector of the cell boundary, and W is the test function.

We assume that the approximate distribution $U_i^h(\mathbf{x}, t)$ inside the cell is a k-th order polynomial, which can be expressed on a series of polynomial bases $\phi_i(x, y) = \{\varphi_{l,i}(x, y)\}_{l=0,\ldots,K}$,

$$U_i^h(\mathbf{x}, t) = \sum_{l=0}^{K} U_i^l(t) \varphi_{l,i}(\mathbf{x}), \tag{5.3}$$

where $K = (k+1)(k+2)/2 - 1$, U_i^l denotes the degree of freedom (DOF). To reduce the stiffness of the discretization equations, the local hierarchical orthogonal basis is employed. Furthermore, we require that $\varphi_{0,i} = 1$ and other basis functions are zero mean functions on cell Ω_i. As a result, it is easy to show that U_i^0 is the cell average of U on Ω_i. Following a similar idea of [17], the specific forms of the basis functions are

$$\begin{aligned}
\varphi_{0,i} &= 1, \\
\varphi_{1,i} &= \Delta x, \\
\varphi_{2,i} &= a_{2,0} + a_{2,1} \Delta x + \Delta y, \\
\varphi_{3,i} &= a_{3,0} + a_{3,1} \Delta x + a_{3,2} \Delta y + \Delta x^2, \\
\varphi_{4,i} &= a_{4,0} + a_{4,1} \Delta x + a_{4,2} \Delta y + a_{4,3} \Delta x^2 + \Delta x \Delta y, \\
\varphi_{5,i} &= a_{5,0} + a_{5,1} \Delta x + a_{5,2} \Delta y + a_{5,3} \Delta x^2 + a_{5,4} \Delta x \Delta y + \Delta y^2, \\
&\ldots \\
\Delta x &= \frac{x - x_i}{\sqrt{\|\Omega_i\|}}, \ \Delta y = \frac{y - y_i}{\sqrt{\|\Omega_i\|}},
\end{aligned} \tag{5.4}$$

where (x_i, y_i) and $\| \ \Omega_i \ \|$ are the barycenter and area of cell Ω_i. The coefficients $a_{m,l}$ can be obtained by satisfying the orthogonal property, i.e.,

$$\int_{\Omega_i} \varphi_{m,i} \varphi_{n,i} \mathrm{d}\Omega = w_m \delta_{mn}, \ w_m = \int_{\Omega_i} (\varphi_{m,i})^2 \mathrm{d}\Omega$$

By taking the basis function to be the test function and substituting the approximate solution (5.3) into Eq. (5.2), the DOFs can be updated by

$$\frac{\partial U_i^l}{\partial t} \int_{\Omega_i} (\varphi_{l,i})^2 d\Omega + \oint_{\partial \Omega_i} \widehat{F}_n \cdot \mathbf{n} \varphi_{l,i} ds - \int_{\Omega_i} \mathbf{F}_n(U^h) \cdot \nabla \varphi_{l,i} \mathrm{d}\Omega = 0, l = 0, \cdots, K.$$

The interfacial flux is computed using the approximate Riemann solver, for which the HLL Riemann solver [8] is adopted in this paper. The volume and surface integrals are calculated using $2k$ and $2k+1$ order accurate Gauss quadrature rules. The resulting semidiscrete system is integrated temporally using the third order Runge-Kutta scheme.

5.2 The Calculation of Viscous Terms

In the evaluation of viscous terms in high-order FVM in Sect. 2.4, the gradient on the face is the mean value of the gradients on the left and right cells. This calculations would make the viscous terms one order lower than inviscid terms. Whereas it is inconsistent in the DG scheme [3]. To solve the problem, Bassi and Rebay [1] proposed the so called BR1 method. In this method, the first order term is replaced with an auxiliary variable and the DG framework is used to discretize this equation with the flux calculated using the mean value on the left and right. Cockburn et al. [3] proved the stability and consistency of BR1 scheme and proposed the local discontinuous Galerkin method (LDG) to better improve the accuracy of viscous terms. To make the scheme compact, Bassi and Rebay [2] proposed the improved BR2 scheme which only relies on the information of left and right cells with the local lift operator. In addition, there exist some other methods, e.g., the general Riemann solver using the penalty method [7], the rDG [16] and RDG [14] method using the reconstruction method. We mainly introduce the BR1 and BR2 scheme in the following section.

For the conservative Navier-Stokes equation,

$$\frac{\partial U}{\partial t} + \nabla \cdot \mathbf{F}(U) - \nabla \cdot \mathbf{F}_v(U, \nabla U) = S$$

with the viscous terms $\mathbf{F}_v(U, \nabla U)$, we introduce the auxiliary variable

$$Q = \nabla U. \tag{5.5}$$

Then the original equation can be written as

$$\frac{\partial U}{\partial t} + \nabla \cdot \mathbf{F}(U) - \nabla \cdot \mathbf{F}_v(U, Q) = S. \tag{5.6}$$

The Eqs. (5.5) and (5.6) is equivalence to the original Navier-Stokes equations. But only first order terms exist in these two equations. Similarly, we expand the auxiliary variable Q using the same basis as the conservative variables, i.e.,

$$Q_i(\mathbf{x}) = \sum_{l=0}^{K} Q_i^l \varphi_{l,i}(\mathbf{x}).$$

The DG discretization is used for the Eqs. (5.5) and (5.6),

$$\begin{cases} \frac{\partial U_i^l}{\partial t} \int\limits_{\Omega_i} (\varphi_{l,i})^2 \mathrm{d}\Omega + \oint\limits_{\partial\Omega_i} (\hat{F} - \hat{F}_v) \cdot \mathbf{n}\varphi_{l,i} \mathrm{d}s - \int\limits_{\Omega_i} (\mathbf{F} - \mathbf{F}_v) \cdot \nabla\varphi_{l,i} \mathrm{d}\Omega = 0 \\ Q_i^l \int\limits_{\Omega_i} (\varphi_{l,i})^2 \mathrm{d}\Omega = \oint\limits_{\partial\Omega_i} \hat{U}\mathbf{n}\varphi_{l,i} \mathrm{d}s - \int\limits_{\Omega_i} U\nabla\varphi_{l,i} \mathrm{d}\Omega \end{cases} \tag{5.7}$$

The second equation in (5.7) can give the solution Q_i^l with the flux $\hat{U}$. Similarly, the first equation in (5.7) gives the solution U_i^l with the flux $\hat{F}_v$. So the basic problem is to determine the numerical flux $\hat{U}$ and $\hat{F}_v$, which is also the main difference among the BR1, BR2, and LDG schemes.

According to the characteristic of viscous terms, the direct thought is to average the values on the left and right, i.e.,

$$\begin{cases} \widehat{U} = \frac{1}{2}(U^L + U^R) \\ \widehat{F}_v = \frac{1}{2}\left(F_v(U^L, Q^L) + F_v(U^R, Q^R)\right) \end{cases}$$

This method is called the BR1 scheme. Analysis shows that BR1 requires the information from nonadjacent cells [2] and it is not compact. Bassi et al proposed the BR2 scheme to solve this problem. First is to integrate the second equation in (5.6) by part again and we have

$$\int\limits_{\Omega_i} Q(\mathbf{x})\varphi_{l,i}\mathrm{d}\Omega = \oint\limits_{\partial\Omega_i} (\widehat{U} - U)\mathbf{n}\varphi_{l,i}\mathrm{ds} + \int\limits_{\Omega_i} \nabla U\varphi_{l,i}\mathrm{d}\Omega \tag{5.8}$$

Define a local lift operator $R_i^h(\mathbf{x})$,

$$\int\limits_{\Omega_i} R_i^h(\mathbf{x})\varphi_{l,i}\mathrm{d}\Omega = \oint\limits_{\partial\Omega_i} (\widehat{U} - U)\mathbf{n}\varphi_{l,i}\mathrm{ds}$$

Substitute it into (5.8) and get

$$\int\limits_{\Omega_i} Q(\mathbf{x})\varphi_{l,i}\mathrm{d}\Omega = \int\limits_{\Omega_i} (\nabla U + R_i^h)\varphi_{l,i}\mathrm{d}\Omega$$

Since the basic function is arbitrarily chosen, we have

$$Q(\mathbf{x}) = \nabla U + R_i^h(\mathbf{x}) \tag{5.9}$$

Comparing the (5.5) and (5.9), the difference between the auxiliary variable $Q(\mathbf{x})$ and the gradient of U is the global lift operator $R_i^h(\mathbf{x})$. The reasons why the BR1 is not compact is that the global lift operator $R_i^h(\mathbf{x})$ is not compact. Next a local lift operator is defined

$$\int\limits_{\Omega_i} r_{i,\sigma}^h(\mathbf{x})\varphi_{l,i}\mathrm{d}\Omega = \int\limits_{\sigma} (\hat{U} - U)\mathbf{n}\varphi_{l,i}\mathrm{ds}, \quad \sigma \in \partial\Omega_i$$

to reflect the discontinuity on the face σ. The relations between the local lift and the global lift is that

$$R_i^h(\mathbf{x}) = \sum_{\sigma \in \partial \Omega_i} r_{i,\sigma}^h(\mathbf{x}). \tag{5.10}$$

So far, the viscous flux in BR2 scheme can be written as

$$\widehat{F}_v = \frac{1}{2}\left(F_v(U^L, \nabla U^L + \eta r_{L,\sigma}^h) + F_v(U^R, \nabla U^L + \eta r_{R,\sigma}^h)\right)$$

where the parameter η affects the stability and is chosen to be the number of sides in the control volume. The local lift operator $r_\sigma^h(\mathbf{x})$ only relies on the direct neighbors of the control volume, which will be beneficial to the implementation of implicit time integration.

5.3 The Construction of the Multidimensional Limiters

In this section, a multidimensional limiting procedure for the DG methods is developed. The proposed limiters are applied on cells containing discontinuities to remove the nonphysical oscillations associated with the high-order DG schemes. Compared to the FV methods, DG methods have more DOFs inside a cell, which enables DG to have compact stencil and to achieve higher resolutions. The most challenging task in designing limiters for the DG method is to keep all the merits of the DG methods. Therefore, the limiting procedure for the DG methods should be designed judiciously so that the limiters do not significantly impair the performance of the DG schemes. The present limiting procedure consists of two steps: (1) several candidate polynomials are reconstructed using the SR approach; and (2) all candidate polynomials are weighted averaged using the WENO procedure. These two steps are described in the following subsections.

5.3.1 The Secondary Reconstruction

The SR is a technique for constructing the candidate polynomials needed in the weighted average procedure of the WENO approach. Before presenting the details of implementation of this technique for the DG methods, the SR for the FV method [10, 11] is reviewed briefly.

In the FV methods of [10, 11], a k-exact reconstruction is first performed on each cell, which is called the primary reconstruction (PR). To provide sufficient number of candidate polynomials for the WENO [10] or the WBAP [11] procedures, the SR is carried out on several control volumes (the stencil for the SR) neighboring the current cell of interest. The basic idea of the SR is to use the continuations of the k-exact PRs on the SR stencil cells as the additional candidate reconstruction polynomials for the current cell. The continuation procedure is managed to guarantee the polynomials obtained from SR that are conservative on the current cell. In [11], the k-exact property of the primary construction is used to show that the SR is accuracy preserving. Furthermore, the polynomials of the SR can be obtained analytically and therefore the SR procedure is very efficient. The stencil of the SR in the k-exact WENO approach [10] consists of the shaded control volumes shown in Fig. 3.1. On the cell of interest, the polynomials of the PR and SRs are used as the candidate polynomials of the WENO procedure.

When the SR technique is used in the DG methods, it must be proceeded with due caution to make the procedure accuracy preserving. To this end, the L_2 projection technique is used to carry out the SR. Assuming cell Ω_j is a neighbor of cell Ω_i, we define the polynomial of SR on cell Ω_i in terms of the information of cell Ω_j as $u_{i,j}(\mathbf{x})$. $u_{i,j}(\mathbf{x})$ is constructed by solving

$$\begin{aligned} &\int_{\Omega_i} u_{i,j}(\mathbf{x})\mathrm{d}\Omega = u_i^0 \left\|\Omega_i\right\|, \\ &\int_{\Omega_j} u_{i,j}(\mathbf{x})\varphi_{l,j}\mathrm{d}\Omega = \int_{\Omega_j} u_j(\mathbf{x})\varphi_{l,j}\mathrm{d}\Omega = u_j^l \int_{\Omega_j} (\varphi_{l,j})^2\mathrm{d}\Omega,\ l = 1, \ldots, K, \end{aligned} \tag{5.11}$$

where u_i^0 is the cell average of the solution on cell Ω_i.

This technique is similar to that used in the reconstruction procedure of the $P_N P_M$ method [6] which has been shown to be accuracy preserving. Moreover, the present SR technique is very simple and efficient. In fact, to obtain the unknowns $u_{i,j}(\mathbf{x})$, there is no need to implement the integration and solve the linear equations. Following a similar idea in [11], we first write $u_{i,j}(\mathbf{x})$ using the orthogonal basis of Ω_i, i.e.,

$$u_{i,j}(\mathbf{x}) = u_i^0 + \sum_{l=1}^{K} u_{i,j}^l \varphi_{l,i}(\mathbf{x}). \tag{5.12}$$

It is obvious that the first equation in Eq. (5.11) is automatically satisfied because of the zero mean property of the basic functions. According to the construction procedure of the hierarchical orthogonal basis, the expression of $\varphi_{l,i}$ can be expressed only by the basis $\varphi_{k,j}, k = 0, \ldots, l$, namely $\varphi_{l,i}(\mathbf{x}) = \sum_{k=0}^{l} c_{k,l}\varphi_{k,j}(\mathbf{x})$. Substituting it into Eq. (5.12), we have

$$\begin{aligned} u_{i,j}(\mathbf{x}) &= u_i^0 + \sum_{l=1}^{K} \left(u_{i,j}^l \sum_{k=0}^{l} c_{k,l}\varphi_{k,j}(\mathbf{x}) \right) \\ &= \left(u_i^0 + \sum_{l=1}^{K} u_{i,j}^l c_{0,l} \right) + \sum_{k=1}^{K} \left(\sum_{l=k}^{K} u_{i,j}^l c_{k,l} \right) \varphi_{k,j}(\mathbf{x}). \end{aligned}$$

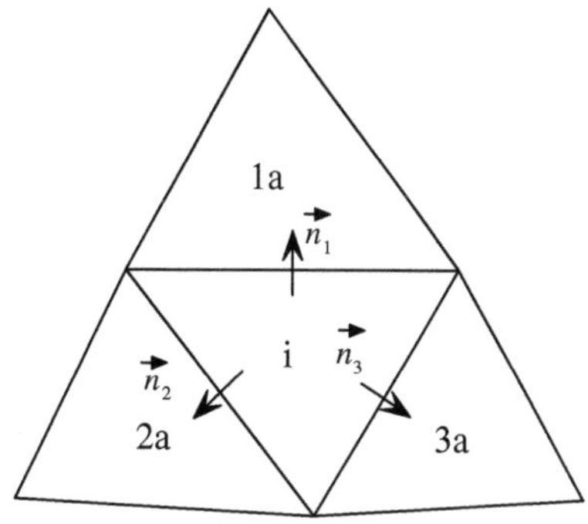

Fig. 5.1 The SR stencil for DG method. $T_i = \{1a, 2a, 3a\}$. Reprinted from Ref. [12], Copyright 2013, with permission from Elsevier

According to the second equation in Eq. (5.11),

$$\sum_{l=k}^{K} u_{i,j}^{l} c_{k,l} = u_j^k, \ k = 1, \cdots, K. \tag{5.13}$$

Using Eq. (5.13), $u_{i,j}^K$ can be computed directly. For $k = K-1, \ldots, 1$ the unknowns $u_{i,j}^k$ are solved recursively in terms of $u_{i,j}^l$, $l = K, \ldots, k+1$. Therefore, the candidate polynomial $u_{i,j}(\mathbf{x})$ of SR can be obtained very efficiently.

For the DG method, it is important to maintain the compactness of the stencil in the limiting procedure. Therefore, unlike the k-exact WENO approach in which the topologically farthest cells in the PR stencil are chosen as the SR stencil of the central cell, the SR stencil of the DG method is chosen as the face neighbors of current cell, which is shown in Fig. 5.1. Therefore, for cell Ω_i, we have three SRs denoted by $u_{i,j}(\mathbf{x})$, $j \in T_i = \{1a, 2a, 3a\}$.

5.3.2 The WENO Limiting Procedure in Characteristic Space

According to the above procedure, four candidate polynomials can be obtained for each cell, namely $u_i(\mathbf{x}), u_{i,j}(\mathbf{x}), j \in T_i = \{1a, 2a, 3a\}$. The next step is applying WENO procedure to the four polynomials. To ease the presentation below, these polynomials are denoted as $u_{i,k}(\mathbf{x}), k = 0, \cdots, N_L, N_L = 3$, where $u_{i,0}(\mathbf{x})$ represents the polynomial $u_i(\mathbf{x})$, and $u_{i,k}(\mathbf{x}), k = 1, \ldots, N_L$ represents the polynomials $u_{i,j}(\mathbf{x}), j \in T_i$. The final reconstruction polynomial are obtained by the weighted average of these polynomials. Specifically, we perform the weighted procedure using the characteristic variables in the following steps.

(I) For a specified normal direction $\mathbf{n}_m$ at the m-th side shown in Fig. 5.1, the two polynomials on the central cell and the cell sharing the common m-th side, i.e.,

$u_{i,l}(\mathbf{x}), l = 0, m$, are first transformed to characteristic space and the corresponding characteristic polynomials are denoted with $V_l^m(\mathbf{x})$, using

$$V_l^m = (R_i^m)^{-1} U_{i,l}, l = 0, m. \tag{5.14}$$

where R_i^m is the matrix of right eigenvectors of the Jacobian matrix of Eq. (5.1) associated with the direction $\mathbf{n}_{m,i}$ in cell i depicted in Fig. 5.1, m is the index of the direction.

(II) Compute the smoothness indicators using

$$\beta_l = \sum_{p+q=1}^{K} h^{p+q-1} \iint_{\Omega_i} \left(\frac{\partial^{p+q} v_l^m(\mathbf{x})}{\partial x^p \partial y^q} \right)^2 \mathrm{d}x\mathrm{d}y, l = 0, m,$$

where v_l^m is the component of V_l^m and h is the length scale of cell Ω_i.

(III) In terms of these smoothness indicators, the WENO weight of each candidate polynomial is determined by

$$w_l = \frac{\alpha_l}{\alpha_0 + \alpha_m}, \quad \alpha_l = \frac{\lambda_l}{(\varepsilon + \beta_l)^p}, l = 0, m$$

Here we choose $p = 2$. $\varepsilon = 10^{-16}$ is a small positive number to avoid possible division by zero, and λ_l is the linear weight. Following the same reason of [4], it is beneficial to emphasize the contribution of the primary polynomial. Therefore, λ_l is computed by

$$\lambda_l = \begin{cases} \lambda_0, & l = 0, \\ 1, & \text{else}. \end{cases}$$

λ_0 can be chosen as 10^3 and 10^5 in our computation. The emphasis of the central weight makes the present approach does not adversely affect the subsell resolution of the DG methods significantly. Larger λ_0 will be beneficial for the accuracy, while smaller λ_0 can suppress the oscillations efficiently. In Sect. 5.4, we give numerical examples on the influence of this parameter. The final polynomial for the cell Ω_i would be the weighted summation of the two candidate polynomials,

$$\tilde{v}_i^m(\mathbf{x}) = w_0 \cdot v_0^m(\mathbf{x}) + w_m \cdot v_m^m(\mathbf{x}).$$

(IV) The steps (I), (II), (III) are carried out for each direction of the sides, e.g., for the triangular control volume, $m = 1, 2, 3$. The already limited polynomials using characteristic variables, $\tilde{v}_i^m(\mathbf{x})$, are then transformed back to the conservative variables which are accordingly denoted by $\tilde{u}_i^m(\mathbf{x})$. Therefore, we obtain three sets of limited polynomials associated with the three directions. From the point of view of practical applications, it is better to have a unique limited polynomial on each cell to remove the ambiguousness in computing the source terms of the DG discretization.

In the present chapter, $\tilde{u}_i^m(\mathbf{x})$, $m = 1, 2, 3$ are further limited using

$$\tilde{u}_i(\mathbf{x}) = \sum_{m=1}^{3} w_{\mathbf{n}_m} \tilde{u}_i^m(\mathbf{x})$$

where $w_{\mathbf{n}_m}$ is the WENO weight for $\tilde{u}_i^m(\mathbf{x})$ in the m-th directions. And $\tilde{u}_i(\mathbf{x})$ is the final limited polynomial. In this part, the WENO weights are computed by

$$w_{\mathbf{n}_m} = \frac{\alpha_{\mathbf{n}_m}}{\sum\limits_{k}^{3} \alpha_{\mathbf{n}_k}}, \; \alpha_{\mathbf{n}_m} = \frac{1}{\left(\varepsilon + \beta_{\mathbf{n}_m}\right)}, m = 1, 2, 3,$$

where the smooth indicator $\beta_{\mathbf{n}_m}$ is evaluated by

$$\beta_{\mathbf{n}_m} = \sum_{p+q=1}^{K} h^{p+q-1} \iint\limits_{\Omega_i} \left(\frac{\partial^{p+q} \tilde{u}_i^m(\mathbf{x})}{\partial x^p \partial y^q} \right)^2 \mathrm{d}x\mathrm{d}y, \quad m = 1, 2, 3.$$

The linear weight for calculating $\alpha_{\mathbf{n}_m}$ is set to be 1 as we treat every direction equally.

The limiting procedure proposed in this section only uses information of face neighbor cells, and thus keeps the compactness property of DG methods. Comparing to other limiting procedures in the characteristic space, e.g. [5, 11], the present characteristic limiting procedure is considerably more efficient. For a specific characteristic direction in the k-th face, the methods in [5, 11] need to transform all the candidate polynomials to ones using characteristic variables, while the methods described above only involve the candidate polynomials on the central cell and the cell sharing the k-th face. Therefore, for a M-face control volume, the present method will be $(M + 1)/2$ times more efficient than that in [5]. According to the numerical examples shown in Sect. 5.4, it can achieve high-order accuracy in smooth region and suppress the oscillations near the discontinuities.

Since the discontinuities are local phenomenon in the flow field, the shock detector is used in some test cases. In this chapter, the parameter free KXRCF shock detector [9] is adopted.

5.4 Numerical Tests

In this section, we apply the DG methods with the multidimensional limiters proposed in the previous sections to solve a number of 2D test cases. These test cases are used to test the accuracy, robustness, resolution, and shock capturing capability of the proposed schemes. In all test cases, the CFL numbers are taken as 80% of the maximum allowable values obtained from the linear stability considerations.

Table 5.1 Accuracy comparison between unlimited DG and DG with WENO limiter. Reprinted from Ref. [12], Copyright 2013, with permission from Elsevier

	Scheme	Grid	L_1 error	Order	L_∞ error	Order
Unlimited DG	Second-order	20	8.04E-04		1.59E-02	
		40	1.79E-04	2.17	4.42E-03	1.85
		80	4.08E-05	2.13	1.21E-03	1.87
		160	9.85E-06	2.05	2.86E-04	2.08
	Third order	20	7.32E-05		1.70E-03	
		40	8.10E-06	3.17	2.31E-04	2.88
		80	8.49E-07	3.25	3.50E-05	2.73
		160	1.07E-07	2.99	4.06E-06	3.10
WENO limiter	Second order	20	1.16E-03		2.86E-02	
		40	2.13E-04	2.45	5.97E-03	2.26
		80	4.30E-05	2.31	1.18E-03	2.33
		160	1.00E-05	2.10	2.94E-04	2.01
	Third order	20	2.38E-04		4.45E-03	
		40	2.11E-05	3.49	5.05E-04	3.14
		80	1.33E-06	3.99	5.24E-05	3.27
		160	1.80E-07	2.88	1.09E-05	2.27

There are some test cases for which the solutions are one-dimensional in nature. They are nevertheless solved on two-dimensional domains. For the problems with curved boundaries, we use the same method in Sect. 4.3 to construct the high-order distribution for the curved boundaries and carry out the quadratures in DGM using isoparametric transformation.

5.4.1 Accuracy Tests

5.4.1.1 Isentropic Vortex

The problem description and grid setup can be referred to Sect. 3.8.1. We compute the solution at $t = 2.0$ to test the accuracy of the proposed schemes without any shock detector. The grids are the same as those used in [10]. The numerical results are shown in Table 5.1 for the unlimited DG scheme and DG scheme with limiters. For the WENO limiter, λ_0 is set to be 10^3 and no shock detector is used. In terms of convergence rate of the density, all schemes can achieve their theoretical values in terms of both L_1 and L_∞ norm. From both Table 5.1 and Fig. 5.2, we can find that the errors for second order DG scheme with limiter are only a bit larger than those for the unlimited DG, while errors using third order DG scheme with limiter are about twice as large as those for the unlimited scheme. Overall the DG schemes with limiters keep the orders unchanged but are more dissipative than the unlimited one.

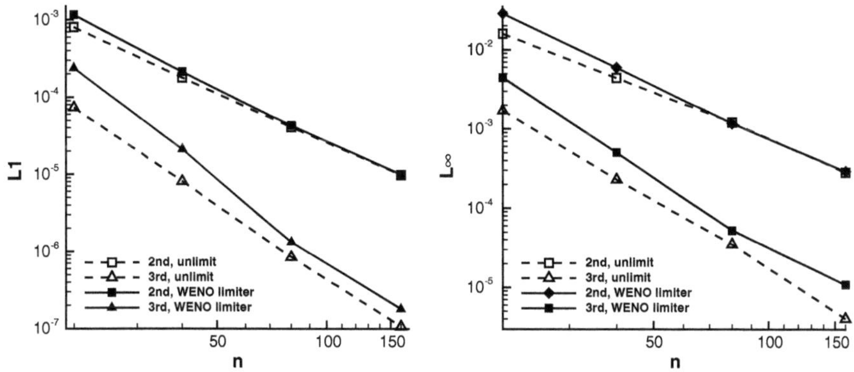

Fig. 5.2 Accuracy for isentropic vortex problem for DG method with/without limiter

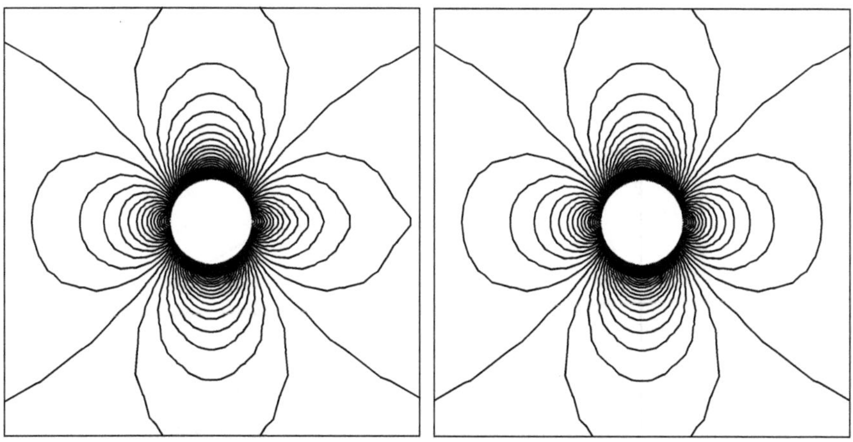

Fig. 5.3 Computed Mach number contours obtained by the second order (*left*) and third order (*right*) DG methods with WENO limiters on 64 × 17 grids

5.4.1.2 Subsonic Flows Past a Circular Cylinder

This case is to calculate the subsonic flows past a circular cylinder at a Mach number of $M_\infty = 0.38$. It is chosen to validate the accuracy of the numerical schemes in solving problems with curved boundaries. Four well-defined O-type grids are used in the computation which are constructed detailedly in [13]. The grids consist of 16×5, 32×9, 64×17 and 128×33 grid points. Computations are performed using second and third order DG schemes with/without limiters on all these four grids. For the DG methods with limiters, λ_0 is set to be 10^3 and no shock detector is used.

The Mach contours obtained using the second order and third order DG methods with WENO limiters on 64 × 17 grids are shown in Fig. 5.3. One can see that the solution obtained by second order WENO limiter is more dissipative than that using third order WENO limiter. To assess the performance of the schemes, we measure

Table 5.2 Accuracy comparison using subsonic flows past a circular cylinder. Reprinted from Ref. [12], Copyright 2013, with permission from Elsevier

Mesh	L_2 error	Order	L_2 error	Order
	Second order without limiter		*Second order WENO limiter*	
16 × 5	2.04E-03		2.61E-03	
32 × 9	4.36E-04	2.23	4.20E-04	2.63
64 × 17	7.27E-05	2.59	5.24E-05	3.00
128 × 33	1.15E-05	2.66	7.79E-06	2.75
	Second order without limiter		*Third order WENO limiter*	
16 × 5	3.51E-04		4.05E-03	
32 × 9	3.27E-05	3.42	1.82E-04	4.47
64 × 17	3.33E-06	3.30	2.19E-05	3.06
128 × 33	4.30E-07	2.95	1.33E-06	4.03

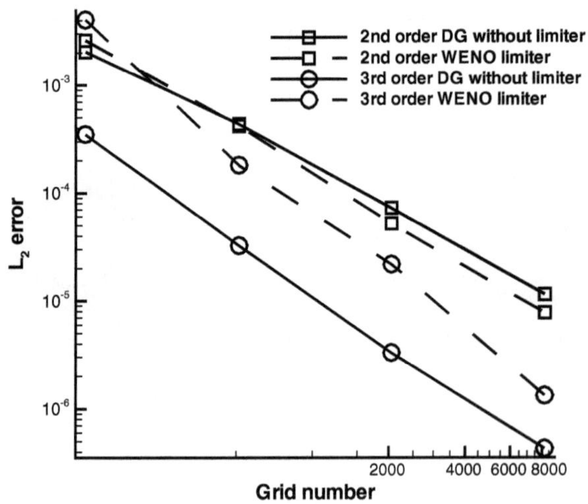

Fig. 5.4 Results of grid-refinement study for flow past a cylinder obtained using DG methods with and without limiters

the L_2 errors in entropy computed as

$$\varepsilon_{\text{ent}} = \frac{P/P_\infty}{(\rho/\rho_\infty)^\gamma} - 1,$$

where P_∞ and ρ_∞ are pressure and density of the free stream. The results of the errors using different schemes are shown in Table 5.2 and Fig. 5.4. From the table and figure, we found the L_2 errors of second order DG with WENO limiters are even smaller than those without limiters, while the errors of the third order DG with WENO limiters are a bit larger than those without limiters. Anyway, all the schemes can achieve the expected orders.

5.4.2 Shock Capturing Capability

5.4.2.1 Linear Advection

The first two-dimensional test case on unstructured grids is the initial value problem for the linear advection equation

$$\begin{aligned} &u_t + u_x + u_y = 0, 0 \le x \le 2, t > 0 \\ &u(x, y, 0) = u_0(x, y), \end{aligned}$$

with the following initial condition:

$$u_0(x) = \begin{cases} \frac{1}{6}\left(G(x,\beta,z-\delta) + G(x,\beta,z+\delta) + 4G(x,\beta,z)\right), & 0.2 \le x \le 0.4, \\ 1, & 0.6 \le x \le 0.8, \\ 1 - \left|10(x-0.1)\right|, & 1 \le x \le 1.2, \\ \frac{1}{6}\left(F(x,\beta,z-\delta) + F(x,\beta,z+\delta) + 4F(x,\beta,z)\right), & 1.4 \le x \le 1.6, \\ 0, & \text{otherwise} \end{cases}$$

$$G(x,\beta,z) = e^{-\beta(x-z)^2}$$

$$F(x,\alpha,a) = \sqrt{\max\left(1 - \alpha^2(x-a)^2, 0\right)}$$

where $a = 0.5, z = -0.7, \delta = 0.005, \alpha = 10$, and $\beta = \frac{\log 2}{36\delta^2}$. The initial profile consists of the combination of a Gaussian wave, a square pulse, a sharp triangle, and a half-ellipse. The computational domain is taken as $[0, 2] \times [0, 0.4]$, and periodic boundary conditions are used. The grids used for computation are uniform triangular grids with grid size $h = 1/100$. The solution is computed to $T = 16$ using second and third order DG methods with different λ_0. Figure 5.5 compares the performance of DG methods with different λ_0. For both $\lambda = 1000$ and 10000, the second and third order DG methods can capture the discontinuities without oscillations. Larger λ_0 and high-order schemes can greatly increase the accuracy.

5.4.2.2 Lax Shock Tube Problem

The Lax tube problem is solved to test the nonoscillation property of the second and third order DG schemes. The initial data are

$$(\rho, u, p)(\mathbf{x}) = \begin{cases} (\rho_L, u_L, p_L) & \text{if } x \le 0.5 \\ (\rho_R, u_R, p_R) & \text{if } x > 0.5 \end{cases}$$

where the left and right states are

$$\begin{aligned} (\rho_L, u_L, p_L) &= (0.445, 0.698, 3.528), \\ (\rho_R, u_R, p_R) &= (0.5, 0., 0.571). \end{aligned}$$

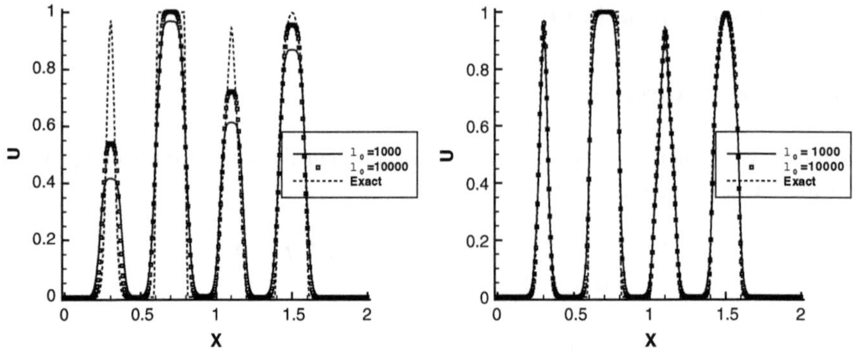

Fig. 5.5 Solution for linear advection problem at $T = 16$. *Left* second order; *right* third order

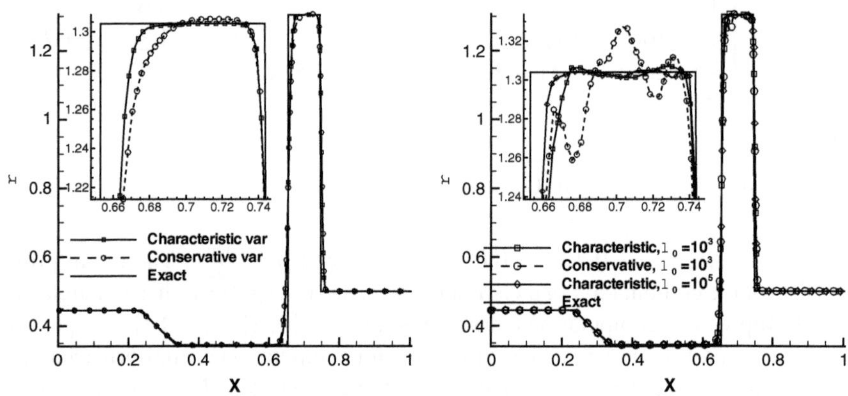

Fig. 5.6 Solution for lax tube problem. *Left* second order; *right* third order

The computational domain is $[0, 1] \times [0, 0.2]$ filled by uniform triangular grids the same as Section 5.4.2.1 with grid size $h = 1/200$. We output the solution at $T = 0.1$. As shown in the first subfigure in Fig. 5.6, no oscillations occur near the shock or contact discontinuity for the second DG scheme with limiters, and the limiters using the characteristic variables can greatly increase the resolutions near the contact discontinuities. From the second subfigure in Fig. 5.6 using the third order DG scheme with limiters, great noises are produced using the conservatives variables for limiting, while few noises are observed in the smooth region using the characteristic variables with parameter $\lambda_0 = 10^3$ or even 10^5.

5.4.3 Double Mach Reflection

The problem description and grid setup can be referred to in Sect. 3.9.4.

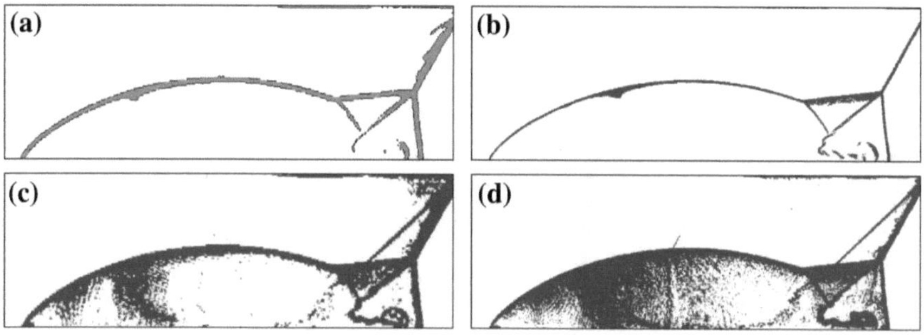

Fig. 5.7 Shock region (*marked in black*) of double Mach reflection detected by the KXRCF shock detector. **a** 2nd order, h=1/120 **b** 2nd order, h=1/240 **c** 3rd order, h=1/120 **d** 3rd order, h=1/240. Reprinted from Ref. [12], Copyright 2013, with permission from Elsevier

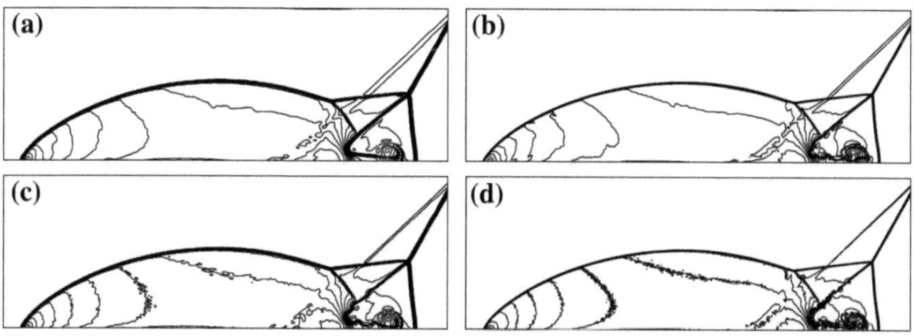

Fig. 5.8 Comparison of density contours for double Mach reflection. Thirty equally spaced contour lines from ρ =2.0 to 21.7. **a** 2nd order, h=1/120 **b** 2nd order, h=1/240 **c** 3rd order, h=1/120 **d** 3rd order, h=1/240. Reprinted from Ref. [12], Copyright 2013, with permission from Elsevier

We compute the result on a coarse grid ($h = 1/120$) and a fine grid ($h = 1/240$) using second and third order DG schemes. The shock regions detected by the KXRCF shock detector are shown in Fig. 5.7 and lots of cells are detected mistakenly as the trouble cells especially for the third order DG schemes. This indicates that the shock detector needs to be improved further. Nevertheless, since the present WENO limiter is accuracy preserving, there is no problem to use it in the smooth regions. According to Fig. 5.8, no oscillations are observed in the results computed using second order limiters and only small noises are found in the results computed using third order limiters. From the slip lines shown in Fig. 5.9, we can clearly see that the third order scheme can capture more detailed flow structures than the second order scheme.

5.4.4 Mach 3 Wind Tunnel with a Step

The problem description and grid setup can be referred to in Sect. 3.9.5. The numerical solutions are computed on both a coarse grid ($h = 1/80$) and a fine grid ($h = 1/160$)

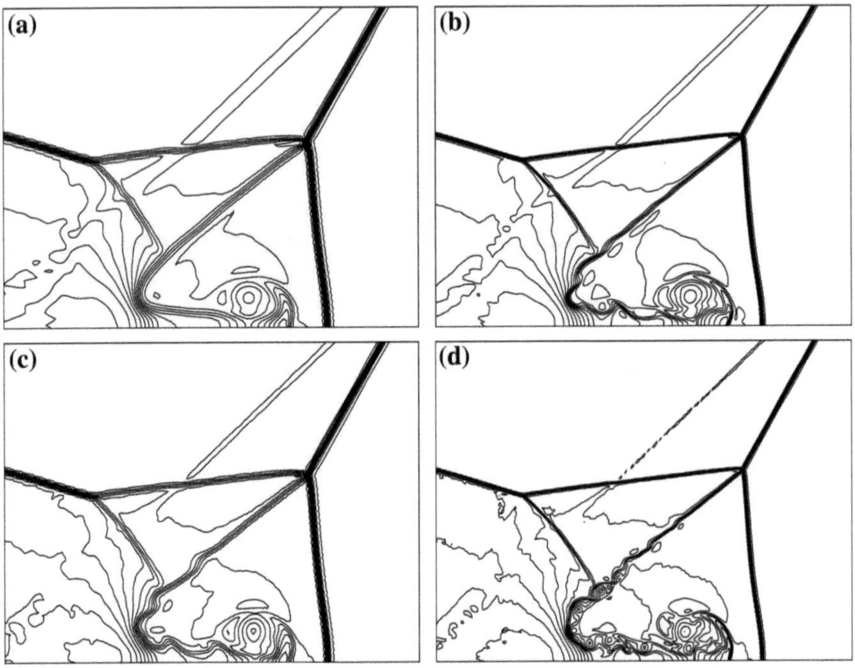

Fig. 5.9 Close-up view around the double Mach stem of Fig. 5.8. **a** 2nd order, h=1/120 **b** 2nd order, h=1/240 **c** 3rd order, h=1/120 **d** 3rd order, h=1/240. Reprinted from Ref. [12], Copyright 2013, with permission from Elsevier

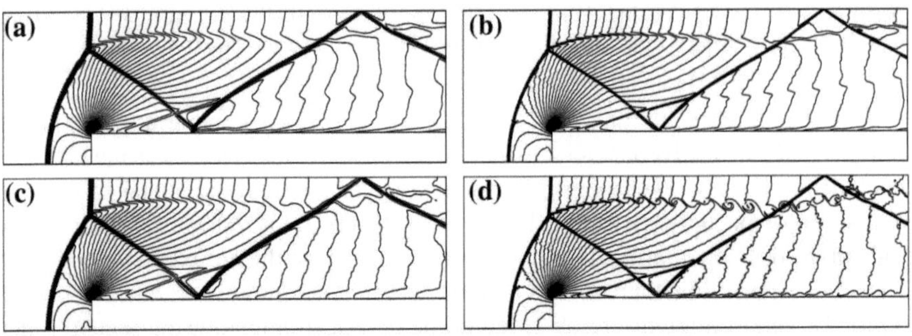

Fig. 5.10 Comparison of density contours for the Mach 3 wind tunnel with a step. Thirty equally spaced contour lines from ρ =0.23 to 6.1. **a** 2nd order, h=1/180 **b** 2nd order, h=1/160 **c** 3rd order, h=1/80 **d** 3rd order, h=1/160. Reprinted from Ref. [12], Copyright 2013, with permission from Elsevier

with the second and third order DG schemes. The results in terms of the density contours are presented in Fig. 5.10. We can see that third order scheme can resolve the shear layer instability better than second order scheme.

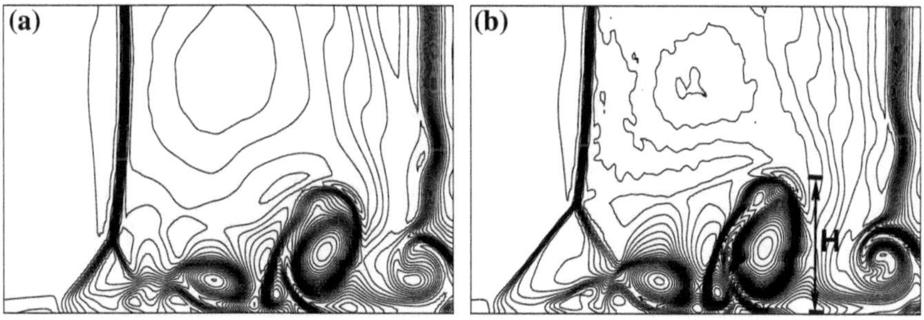

Fig. 5.11 The density contour of the viscous shock tube using the second and third order DG schemes. **a** 2nd order **b** 3rd order

Table 5.3 Vortex height comparison using different schemes

Vortex height	Second order	Third order	Fourth order	Reference
FVM	0.12	0.145	0.16	0.169
DGM	0.15	0.165		

5.4.5 Viscous Shock Tube Problem

The problem description and grid setup can be referred to in Sect. 3.9.8. In this section the second and third order DG schemes are used in the computation. HLL scheme is used as the Riemann solver and BR2 scheme is used to treat the viscous terms. As shown in Fig. 5.11, the second and third order DG can capture the shock waves and vortex in the flow field. With higher order, the scheme can capture more details. To compare the resolution between different schemes, we measure the height for the vortex in Fig. 5.11, and the results are shown in Table 5.3. We can find that the results using the DG schemes come closer to reference ones than that using one-order higher FVM.

5.4.6 The Unsteady Flow Around the Sphere

This test case is to validate the curve boundary and viscous terms treatment in three-dimensional DG schemes. The problem description is seem in Sect. 4.4.6. This section calculates the problem using the third order DG scheme. Since there are 10 DOFs in every cell, the grid size can be chosen two times that of FVM. Thus tetrahedral grids consisting of 371,752 cells and 63,083 grid points are used in the calculation with grid size $h = 0.3$ near the wall boundary.

Similarly, the Q-criteria is used to identify the vortex as shown in Fig. 5.12. Since we only output one DOF in each cell, the contour is not smooth. The mach number

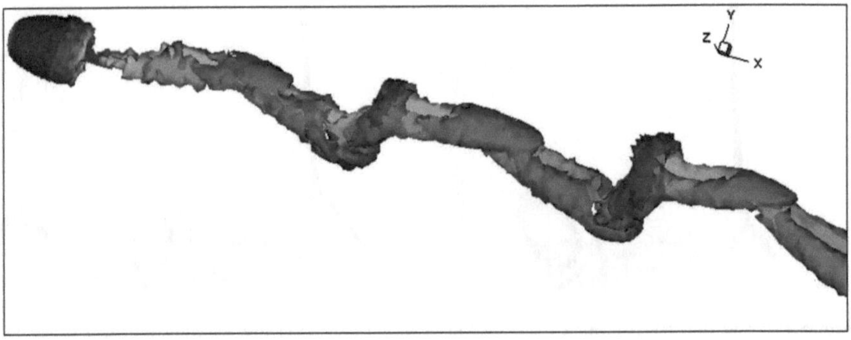

Fig. 5.12 The vortex using Q-criteria at $Q = 0.001$ colored by Mach number

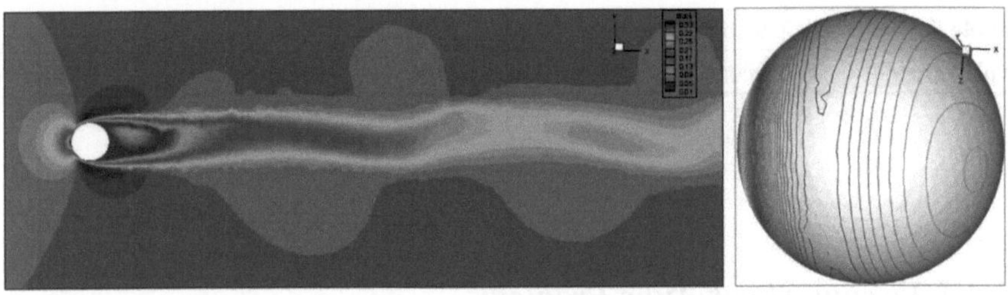

Fig. 5.13 The contour of Mach number in the wake at the section $z = 0$ (*left*) and the pressure contour (*right*)

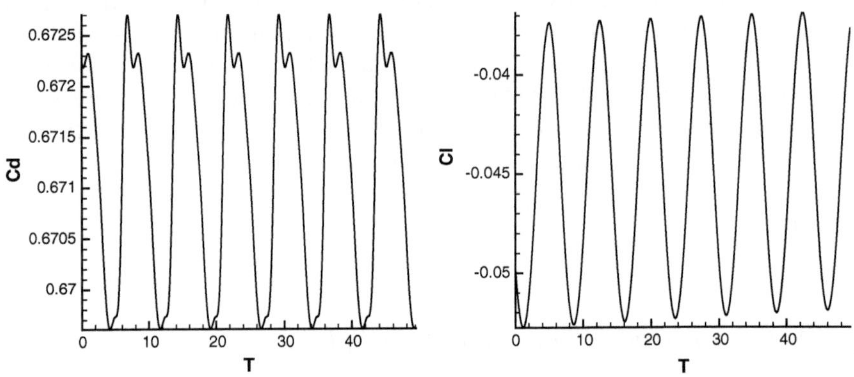

Fig. 5.14 The history of lift (*left*) and drag (*right*) coefficients

contour shown on the left of Fig. 5.13 is similar to that of finite volume scheme in Sect. 4.4.6. The smooth pressure on the surface on Fig. 5.13 validates the correctness of curved boundary treatment. Through the history of lift and drag coefficients shown in Fig. 5.14, the Strouhal number is evaluated as $St = fd/ = 0.132$ which is close to 0.133 in Sect. 4.4.6.

5.5 Conclusions

In this chapter, a new limiter based on the secondary reconstruction and the WENO approach is developed for the discontinuous Galerkin method. An efficient secondary reconstruction technique is proposed which provides the candidate polynomials used in the weighted average procedure of the WENO approach. The secondary reconstructions are performed only on the face neighboring cells to keep the compactness of the discontinuous Galerkin method. The central weight of the WENO approach is emphasized to maintain the resolution of the DG methods. A new characteristic limiting way is put forward to efficiently eliminate the oscillations near the shock waves in nonlinear equations on unstructured grids. Numerical examples are provided to show that the limiters are robust enough to suppress the oscillations near the shock and are accuracy preserving in smooth flow filed. Further work will be carried out to improve the limiters to work on arbitrary high-order schemes and other types of grids.

References

1. Bassi F, Rebay S (1997) A high-order accurate discontinuous finite element method for the numerical solution of the compressible Navier-Stokes equations. J Comput Phys 131:267–279
2. Bassi F, Crivellini A, Rebay S et al (2005) Discontinuous Galerkin solution of the Reynolds-averaged Navier-Stokes and turbulence model equations. Comput Fluids 34:507–540
3. Cockburn B, Shu CW (2001) Runge-Kutta discontinuous Galerkin methods for convection-dominated problems. J Sci Comput 16(3):173–261
4. Dumbser M, Käser M (2007) Arbitrary high order non-oscillatory finite volume schemes on unstructured meshes for linear hyperbolic systems. J Comput Phys 221:693–723
5. Dumbser M, Käser M, Titarev VA, Toro EF (2007) Quadrature-free non-oscillatory finite volume schemes on unstructured meshes for nonlinear hyperbolic systems. J Comput Phys 226:204–243
6. Dumbser M (2010) Arbitrary high order PNPM schemes on unstructred meshes for the compressible Navier-Stokes equations. Comput Fluids 39:60–76
7. Gassner G, Lörcher F, Munz CD (2007) A contribution to the construction of diffusion flues for finite volume and discontinuous Galerkin schemes. J Comput Phys 224:1049–1063
8. Harten A, Lax P, van Leer B (1983) On upstream differencing and Godunov type methods for hyperbolic conservation laws. SIAM Rev 25:35–61
9. Krivodonova L, Xin J, Remacle J-F, Chevaugeon N, Flaherty JE (2004) Shock detection and limiting with discontinuous Galerkin methods for hyperbolic conservation laws. Appl Numer Math 48:323–338
10. Li W, Ren Y-X ((2012)) High order k-exact WENO finite volume schemes for solving gas dynamic Euler equations on unstructured grids. Int J Numer Meth Fluids 70:742–763
11. Li W, Ren Y-X (2012) The multi-dimensional limiters for solving hyperbolic conservation laws on unstructured grids II: extension to high order finite volume schemes. J Comput Phys 231:4053–4077
12. Li W, Ren Y-X (2013) The multi-dimensional limiters for discontinuous Galerkin method on unstructured grids. Comput Fluids doi:10.1016/j.compfluid.2013.12.008
13. Luo H, Baum JD, Löhner R (2007) A Hermite WENO-based limiter for discontinuous Galerkin method on unstructured grids. J Comput Phys 225:686–713
14. Luo H, Luo L, Nourgaliev R et al (2010) A reconstructed discontinuous Galerkin method for the compressible Navier-Stokes equations on arbitrary grids. J Comput Phys 229:6961–6978

15. Park JS, Kim C (2011) Higher-order discontinuous Galerkin-MLP methods on triangular and tetrahedral grids. In: AIAA 2011–3059, 20th AIAA computational fluid dynamics conference, 27–30 June 2011, Honolulu, Hawaii
16. Van Leer B, Lo MA (2007) Discontinuous Galerkin method for diffusion based on recovery. In: 18th AIAA computational fluid dynamics conference, 25–28 June, Miami, FL
17. Zhu J, Qiu JX, Shu CW, Dumbser M (2008) Runge-Kutta discontinuous Galerkin method using WENO limiters II: unstructured meshes. J Comput Phys 227:4330–4353

Chapter 6
Conclusions and Prospects

This thesis studies the implementation of high order FV and DG methods on unstructured grids. Some novel algorithms are developed to solve series of difficult issues in these two methods. And these algorithms are applied in computing the compressible viscous flows. According to the subject in each chapter, the main work and innovation points can be summarized as follows.

1. Complete the framework of high order unstructured FVM in solving viscous compressible flows. The mature k-exact reconstruction and flux computation methods are used in the implementation of 2D and 3D unstructured FVM code. The main innovations are the simplification of viscous flux evaluation and the treatment of viscous boundary conditions. The computation of viscous flux fully utilizes the existed information of k-exact reconstruction and saves lots of computations. The treatment of viscous boundary conditions is to incorporate the boundary conditions into the k-exact reconstruction through constrained conditions.
2. Propose the secondary reconstruction procedure to construct the candidate polynomials for the limiters. Comparing to the multiple reconstructions for traditional WENO scheme on unstructured grids, SR procedure can reduce the computational cost and storage. Also the implementation of SR is convenient and has no such ambiguous process in the multiple reconstructions. Based on the secondary reconstruction, three limiting strategies are proposed, i.e., the quadrature-free k-exact WENO limiter, the WBAP limiter with successive limiting procedure and the group weighted limiter. The three novel limiters have the following characteristics. They can suppress the oscillations near the discontinuities while keeping the high order accuracy in smooth regions. Sometimes the accuracy will be raised while applying the limiters in smooth regions. The WBAP limiter gives a new general limiting strategy other than existing limiters, which is even more efficient than the improved quadrature-free k-exact WENO limiter in terms of computational cost and implementation convenience. Meanwhile, theoretical analysis indicates that its extreme form can satisfy TVD property or maximum principle under some conditions. This property guarantees that the WBAP limiter can efficiently suppress the numerical oscillations.

W. Li, *Efficient Implementation of High-Order Accurate Numerical Methods on Unstructured Grids*, Springer Theses, DOI: 10.1007/978-3-662-43432-1_6,

3. Use the iso-parametric transformation to generalize the high order FVM to apply on mixed grids and offer some simplified ways for treating curved wall boundaries. The thesis continues to validate the efficiency of the high order reconstruction and the limiting strategies in mixed grids. And on the curved boundaries, the thesis provides some simplified iso-parametric transformation with less control points and simplifies the implement process.
4. Generalize the new limiting strategies and boundary treatment methods to the compact DGM. First is using the secondary reconstruction and WENO limiters to construct a non-oscillatory DG schemes. Second is treating the curved boundaries in DGM with iso-parametric transformation. These methods makes the DGM keep the compactness and high order properties.
5. Integrate the newly developed method into FVM and DGM framework and use the schemes to solve compressible flows including the subsonic and supersonic flows, viscous and inviscid flows. These schemes are validated through accuracy test, the non-oscillatory ability in capturing shock waves, and the resolutions in resolving the boundary layers and vortex. The superiorities of high order schemes are reflected in the computation of wall friction, entropy and vortex structures, where the high order schemes can provide more accurate results than low order schemes.

The unstructured high-order methods are still under continuous development and there still exist some unsolved issues. The ongoing work will involve the following algorithms: (1) The simple and accurate way to represent the three dimensional curved boundaries. (2) The arbitrary high order limiters and implicit time integration schemes for DGM. (3) The *hp* adaptivity in high order schemes. Moreover, the unstructured high-order methods will be used in the simulations of turbulent flows with the direction numerical simulations or large eddy simulations. For the high Reynolds number flows, the RANS turbulent models or detach eddy simulations methods will be incorporated into the high order schemes.

Appendix A
The Derivation of the Two Limits of WBAP Limiters

This chapter will give the derivation of the two limits of WBAP limiters.

For simplification, we do not emphasize the contribution of the central cell, i.e., $\lambda_0 = 1$ in (3.30). In fact, when $\lambda_0 = n$, the WBAP function equals to

$$W(\underbrace{\theta_0, \dots \theta_0}_{n}, \theta_1, \dots, \theta_J).$$

Thus the derivation using $\lambda_0 = 1$ is still valid. For the general form of WBAP function, we can rewrite it into

$$W(\theta_0, \dots, \theta_J) = B^{-1}\left(\frac{\frac{1}{B(\theta_0)} + \dots + \frac{1}{B(\theta_J)}}{\frac{1}{B^2(\theta_0)} + \dots + \frac{1}{B^2(\theta_J)}} \right),$$

where $\theta_0 = 1$. Choose the biased function (3.29), we have

$$W(\theta_0, \dots, \theta_J) = B^{-1}\left(\frac{\sum_{j=0}^{J} \frac{\sqrt{\theta_j^2+\varepsilon^2}}{\theta_j}}{\sum_{j=0}^{J} 1 + \left(\frac{\varepsilon}{\theta_j}\right)^2 \sum_{j=0}^{J}} \right) \tag{A.1}$$

$$= \frac{\varepsilon \sum_{j=0}^{J} \frac{\sqrt{\theta_j^2+\varepsilon^2}}{\theta_j}}{\left(\sum_{j=0}^{J} 1 + \left(\frac{\varepsilon}{\theta_j}\right)^2\right) \sqrt{1 - \left(\frac{\sum_{j=0}^{J} \frac{\sqrt{\theta_j^2+\varepsilon^2}}{\theta_j}}{\sum_{j=0}^{J} 1 + \left(\frac{\varepsilon}{\theta_j}\right)^2} \right)^2}}.$$

W. Li, *Efficient Implementation of High-Order Accurate Numerical Methods on Unstructured Grids*, Springer Theses, DOI: 10.1007/978-3-662-43432-1,

(1) For the first limit $\varepsilon \to 0$, we have

$$W^{L1}(\theta_0, \ldots, \theta_J) = \lim_{\varepsilon \to 0} W(\theta_0, \ldots, \theta_J)$$

$$= \lim_{\varepsilon \to 0} \frac{\sum_{j=0}^{J} \frac{\sqrt{\theta_j^2+\varepsilon^2}}{\theta_j}}{\left(\sum_{j=0}^{J} 1 + \left(\frac{\varepsilon}{\theta_j}\right)^2\right)} \cdot \lim_{\varepsilon \to 0} \sqrt{\frac{\varepsilon^2}{1 - \left(\frac{\sum_{j=0}^{J} \frac{\sqrt{\theta_j^2+\varepsilon^2}}{\theta_j}}{\sum_{j=0}^{J} 1+\left(\frac{\varepsilon}{\theta_j}\right)^2}\right)^2}}.$$

(1.a) When $\theta_0, \ldots, \theta_J > 0$, the above formula is simplified

$$W^{L1}(\theta_0, \ldots, \theta_J)$$

$$= \sqrt{\lim_{\varepsilon \to 0} \left(\frac{\varepsilon^2}{1 - \left(\frac{\sum_{j=0}^{J} \frac{\sqrt{\theta_j^2+\varepsilon^2}}{\theta_j}}{\sum_{j=0}^{J} 1+\left(\frac{\varepsilon}{\theta_j}\right)^2}\right)^2} \right)}$$

$$= \sqrt{\lim_{\varepsilon \to 0} \frac{2\varepsilon}{-2\frac{\sum_{j=0}^{J} \frac{\sqrt{\theta_j^2+\varepsilon^2}}{\theta_j}}{\sum_{j=0}^{J} 1+\left(\frac{\varepsilon}{\theta_j}\right)^2} \cdot \left(\frac{\sum_{j=0}^{J} \frac{\varepsilon}{\theta_j\sqrt{\theta_j^2+\varepsilon^2}}}{\left[\sum_{j=0}^{J} 1+\left(\frac{\varepsilon}{\theta_j}\right)^2\right]} - \frac{\sum_{j=0}^{J} \frac{\sqrt{\theta_j^2+\varepsilon^2}}{\theta_j} \cdot \sum_{j=0}^{J} \frac{2\varepsilon}{\theta_j^2}}{\left[\sum_{j=0}^{J} 1+\left(\frac{\varepsilon}{\theta_j}\right)^2\right]^2} \right)}}$$

$$= \sqrt{\frac{1+J}{\sum_{j=0}^{J} \frac{1}{\theta_j^2}}}.$$

(1.b) When $\forall \theta_j < 0$, $j = 0, \ldots, J$, we notice that the denominator

$$\lim_{\varepsilon \to 0} 1 - \left(\frac{\sum_{j=0}^{J} \frac{\sqrt{\theta_j^2+\varepsilon^2}}{\theta_j}}{\sum_{j=0}^{J} 1 + \left(\frac{\varepsilon}{\theta_j}\right)^2}\right)^2 \neq 0,$$

then $W^{L1}(\theta_0, \ldots, \theta_J) = 0$.

Thus we have the first limit $\mathrm{WBAP} - L1$.

(2) For the second limit $\varepsilon \to \infty$, we have

$$W^{L2}(\theta_0, \ldots, \theta_J) = \lim_{\varepsilon \to \infty} W(\theta_0, \ldots, \theta_J)$$

$$= \lim_{\varepsilon \to \infty} \frac{\varepsilon \sum_{j=0}^{J} \frac{\sqrt{\theta_j^2 + \varepsilon^2}}{\theta_j}}{\left(\sum_{j=0}^{J} 1 + \left(\frac{\varepsilon}{\theta_j}\right)^2\right)} \cdot \lim_{\varepsilon \to \infty} \sqrt{\frac{1}{1 - \left(\frac{\sum_{j=0}^{J} \frac{\sqrt{\theta_j^2+\varepsilon^2}}{\theta_j}}{\sum_{j=0}^{J} 1 + \left(\frac{\varepsilon}{\theta_j}\right)^2}\right)^2}}$$

$$= \lim_{\varepsilon \to \infty} \frac{\varepsilon \sum_{j=0}^{J} \frac{\sqrt{\theta_j^2 + \varepsilon^2}}{\theta_j}}{\left(\sum_{j=0}^{J} 1 + \left(\frac{\varepsilon}{\theta_j}\right)^2\right)} = \lim_{\varepsilon \to \infty} \frac{\sum_{j=0}^{J} \frac{1}{\theta_j}}{\sum_{j=0}^{J} \frac{1}{\theta_j^2}}.$$

Thus we have the second limit WBAP $- L2$.

Appendix B
The Proof of WBAP Lemma 1

Proof The use of W^{L1} in Eqs. (3.26) and (3.27) yields the following limited linear reconstruction

$$\begin{aligned} u_i(x) &= \bar{u}_i + W^{L1}(1, \frac{\bar{\sigma}_{i-1}}{\bar{\sigma}_i}, \frac{\bar{\sigma}_{i+1}}{\bar{\sigma}_i}) \cdot \bar{\sigma}_i \cdot (x - x_i) \\ &= \bar{u}_i + W^{L1}(1, \frac{\delta_{i-1}}{\delta_i}, \frac{\delta_{i+1}}{\delta_i}) \cdot \frac{\delta_i}{2\Delta x} \cdot (x - x_i) \end{aligned}$$

where $\delta_i = 2\Delta x \cdot \bar{\sigma}_i = \bar{u}_{i+1} - \bar{u}_{i-1}$. And the left and right states are given by

$$\begin{aligned} u^-_{i+1/2} &= \bar{u}_i + W^{L1}\left(1, \tfrac{\delta_{i-1}}{\delta_i}, \tfrac{\delta_{i+1}}{\delta_i}\right) \cdot \tfrac{\delta_i}{4} \\ u^+_{i-1/2} &= \bar{u}_i - W^{L1}\left(1, \tfrac{\delta_{i-1}}{\delta_i}, \tfrac{\delta_{i+1}}{\delta_i}\right) \cdot \tfrac{\delta_i}{4}. \end{aligned} \tag{B.1}$$

According to Eq. (3.33), the specific form of W^{L1} is

$$W^{L1}\left(1, \frac{\delta_{i-1}}{\delta_i}, \frac{\delta_{i+1}}{\delta_i}\right) \cdot \frac{\delta_i}{4} = \begin{cases} ll\sqrt{\frac{(n+2)/16}{\frac{n}{\delta_i^2} + \frac{1}{\delta_{i-1}^2} + \frac{1}{\delta_{i+1}^2}}}, & \text{if } \delta_i, \delta_{i+1} \text{ and } \delta_{i-1} > 0 \\ -\sqrt{\frac{(n+2)/16}{\frac{n}{\delta_i^2} + \frac{1}{\delta_{i-1}^2} + \frac{1}{\delta_{i+1}^2}}}, & \text{if } \delta_i, \delta_{i+1} \text{ and } \delta_{i-1} < 0 \\ 0, & \text{otherwise} \end{cases} \tag{B.2}$$

We firstly consider the case δ_i,δ_{i+1} and $\delta_{i-1} > 0$, which is corresponding to the first branch of Eq. (B.2). In this case it is trivial to show that

$$u^-_{i+1/2} \geq \bar{u}_i \geq m, u^+_{i-1/2} \leq \bar{u}_i \leq M. \tag{B.3}$$

Furthermore, if $n \leq 14$, we have

W. Li, *Efficient Implementation of High-Order Accurate Numerical Methods on Unstructured Grids*, Springer Theses, DOI: 10.1007/978-3-662-43432-1,

$$W^{L1}\left(1, \frac{\delta_{i-1}}{\delta_i}, \frac{\delta_{i+1}}{\delta_i}\right) \cdot \frac{\delta_i}{4} \le \sqrt{\frac{1}{\frac{n}{\delta_i^2} + \frac{1}{\delta_{i-1}^2} + \frac{1}{\delta_{i+1}^2}}} \le \delta_{i+1} \tag{B.4}$$

and

$$W^{L1}\left(1, \frac{\delta_{i-1}}{\delta_i}, \frac{\delta_{i+1}}{\delta_i}\right) \cdot \frac{\delta_i}{4} \le \sqrt{\frac{1}{\frac{n}{\delta_i^2} + \frac{1}{\delta_{i-1}^2} + \frac{1}{\delta_{i+1}^2}}} \le \delta_{i-1}. \tag{B.5}$$

In terms of Eq. (B.1),

$$u_{i+1/2}^{-} = \bar{u}_i + W^{L1}\left(1, \frac{\delta_{i-1}}{\delta_i}, \frac{\delta_{i+1}}{\delta_i}\right) \cdot \frac{\delta_i}{4} \le \bar{u}_i + \delta_{i+1} = \bar{u}_{i+2} \le M \tag{B.6}$$

and

$$u_{i-1/2}^{+} = \bar{u}_i - W^{L1}\left(1, \frac{\delta_{i-1}}{\delta_i}, \frac{\delta_{i+1}}{\delta_i}\right) \cdot \frac{\delta_i}{4} \ge \bar{u}_i - \delta_{i-1} = \bar{u}_{i-2} \ge m. \tag{B.7}$$

According to Eqs. (B.3), (B.6) and (B.7), we have

$$u_{i-1/2}^{+}, u_{i+1/2}^{-} \in [m, M].$$

For the case δ_i, δ_{i+1} and $\delta_{i-1} < 0$ which is corresponding to the second branch of Eq. (B.2), the proof can be done in a similar manner. If δ_i, δ_{i+1} and δ_{i-1} are not of the same sign, it is straightforward to show that $m \le u_{i-1/2}^{+} = u_{i+1/2}^{-} = \bar{u}_i \le M$.

MIX
Papier aus verantwortungsvollen Quellen
Paper from responsible sources
FSC® C105338

If you have any concerns about our products,
you can contact us on
ProductSafety@springernature.com

In case Publisher is established outside the EU,
the EU authorized representative is:
Springer Nature Customer Service Center GmbH
Europaplatz 3, 69115 Heidelberg, Germany

Printed by Libri Plureos GmbH
in Hamburg, Germany